Computer Communications and Networks

For other titles published in this series, go to
www.springer.com/series/4198

The Computer Communications and Networks series is a range of textbooks, monographs and handbooks. It sets out to provide students, researchers and non-specialists alike with a sure grounding in current knowledge, together with comprehensible access to the latest developments in computer communications and networking.

Emphasis is placed on clear and explanatory styles that support a tutorial approach, so that even the most complex of topics is presented in a lucid and intelligible manner.

Liljana Gavrilovska · Srdjan Krco
Veljko Milutinovic · Ivan Stojmenovic
Roman Trobec
Editors

Application and Multidisciplinary Aspects of Wireless Sensor Networks

Concepts, Integration, and Case Studies

Springer

Editors
Liljana Gavrilovska
Faculty of EE and IT
Ss. Cyril and Methodius
University of Skopje
Rudjer Boskovic bb, 1000 Skopje
Macedonia
liljana@feit.ukim.edu.mk

Veljko Milutinovic
Department of Computer Engineering
University of Belgrade
Bulevar Kralja Aleksandra 73
11000 Belgrade, Serbia
vm@etf.rs

Roman Trobec
Department of Communication Systems
Jožef Stefan Institute
Jamova 39, 1000 Ljubljana, Slovenia
roman.trobec@ijs.si

Srdjan Krco
Ericsson Software Campus
Ericsson Ireland Research Centre
Athlone
County Westmeath
Ireland
srdjan.krco@ericsson.com

Ivan Stojmenovic
School of IT and Engineering
University of Ottawa
King Edward Ave. 800
K1N 6N5 Ottawa
ON, Canada
stomenovic@storm.ca

Series Editor
Professor A.J. Sammes, BSc, MPhil, PhD,
FBCS, CEng
Centre for Forensic Computing
Cranfield University, DCMT, Shrivenham
Swindon SN6 8LA
UK

ISBN 978-1-4471-2596-9 ISBN 978-1-84996-510-1 (eBook)
DOI 10.1007/978-1-84996-510-1
Springer London Dordrecht Heidelberg New York

British Library Cataloguing in Publication Data
A catalogue record for this book is available from the British Library

Cover design: SPi, Puducherry, India

Printed on acid-free paper

Springer is part of Springer Science+Business Media (www.springer.com)

Foreword: Creativity in Science

Creativity is sought everywhere: in the arts, entertainment, business, mathematics, engineering, medicine, the social sciences, and the physical sciences. Common elements of creativity are originality and imagination. Creativity is intertwined with the freedom to design, to invent, and to dream. In engineering and science, however, creativity is useful only if it fits into the realities of the physical world. A creative idea in science or engineering must conform to the law of conservation of energy (including the mass energy mc^2). An inventor that thinks that she or he knows how to violate the conservation of energy will have to disprove a vast amount of laboratory measurements and accepted theory.

Martin Perl M. Perl
Stanford University, Nobel Laurate

Preface

This edited book is a result of the EU FP7 project ProSense (*Promote, Mobilize, Reinforce and Integrate Wireless Sensor Networking Research and Researchers: Towards Pervasive Networking of West Balkan Countries and the EU, grant agreement 205494)* related to wireless sensor networks (WSN). All the authors of specific chapters in this book have contributed to the project in the period March 2008–March 2010. Working on the project, they had a chance to master the specific domains of this newly emerging field, which enables this book to represent a wholistic coverage of WSN. Chapters are lined up according to the topic they cover, and the authors of the chapters are lined up according to the size of their contribution. The five editors of the book (listed in alphabetical order) are Liljana Gavrilovska, Srdjan Krco, Veljko Milutinovic, Ivan Stojmenovic, and Roman Trobec.

Their major role was to help create the overall structure of the book and to help the talents of the contributing authors to generate their best. They are responsible for the final quality control, having in mind that one of the major purposes of the book is to serve as a support for the educational process at universities. They are also responsible for inducing creativity among young researchers on the project, along the lines expressed by Martin Perl in his foreword.

December 2009

Liljana Gavrilovska
Srdjan Krco
Veljko Milutinovic
Ivan Stojmenovic
Roman Trobec

Contents

Chapter 1
Introduction: Bird's-Eye View of Wireless Sensor Networks

Aleksandar Crnjin

Over the course of the last two decades, sensor networks have emerged from complete obscurity into being an important research subject and are slowly crawling their way into industrial applications. In this chapter, we present a short introduction to sensor networks, from the brief overview of technology, through its applications from the beginning of its history in submarine warfare to diverse civilian applications possible today, to an overview of the issues that sensor network researches and developers are confronted with.

1.1 Overview of Technology

Sensor networks can be viewed as an extreme extrapolation of two general trends in computing: miniaturization (making computers ever smaller) and interconnection (networking) [1]. Individual sensor nodes are small devices, running on limited memory, equipped with microcontrollers used for embedded devices (typically 8 bit, 4–5 MHz). In addition to a microcontroller, a sensor node typically includes a small amount of RAM memory (a few kilobytes), a radio transceiver, a certain number of analog sensors connected to the microcontroller through a A/D convertor circuit, and a power source to power all these components (typically 2 AA batteries; Fig. 1.1).

Individual sensor nodes are not capable of doing very much: they can record some quality of the environment (such as temperature, air pressure, humidity, etc.) but their limited processing ability and limited output capability (in terms of a display, most sensor nodes have just a few LED diodes) mean that sensor nodes need help to process and display the data they collect (Fig. 1.2).

Indeed, sensor nodes are invariably used in *sensor networks* which consist of a large number of sensor nodes, which collect data, and one or more *gateway*

A. Crnjin (✉)
School of Electrical Engineering, University of Belgrade, Serbia
e-mail: acrnjin@gmail.com

L. Gavrilovska et al. (eds.), *Application and Multidisciplinary Aspects of Wireless Sensor Networks*, Computer Communications and Networks,
DOI 10.1007/978-1-84996-510-1_1, © Springer-Verlag London Limited 2011

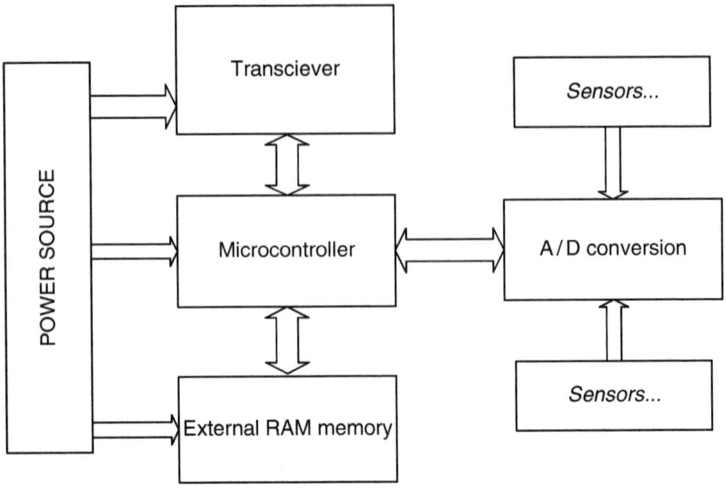

Fig. 1.1 Block diagram of a sensor node

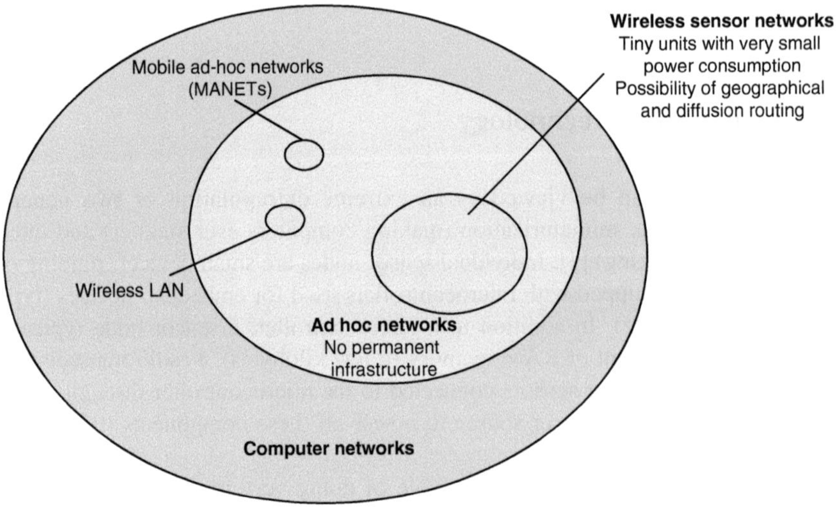

Fig. 1.2 Position of sensor networks in the broad context of computer networking

sensor nodes, which route collected data onward to their eventual destination. For example, a gateway sensor node could be connected to a PC computer, or an Internet router which could relay the data onward to the final destination (Fig. 1.3).

Limited energy resources of sensor nodes (typically, they only run on battery power) and the fact that in large-scale deployment of sensor networks, it would be a major hurdle to replace batteries on nodes which have spent their existing ones, have important implications:

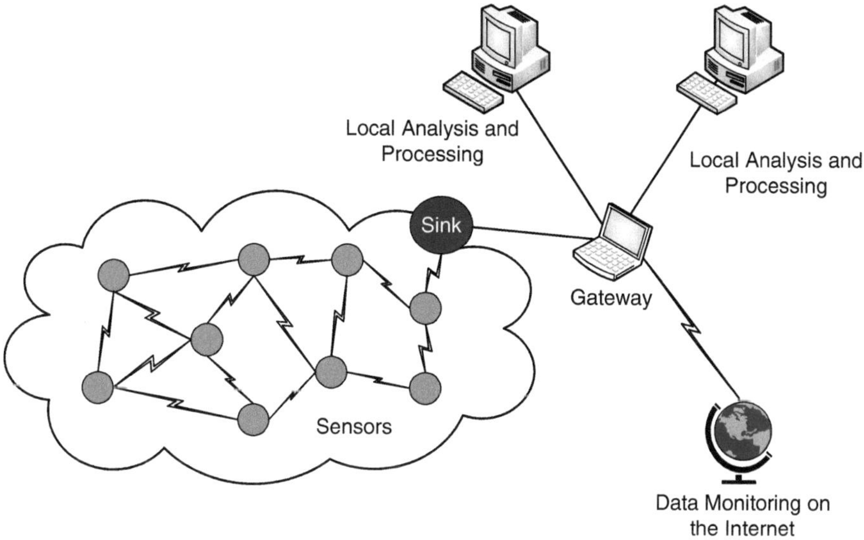

Fig. 1.3 A typical sensor network usage scenario

- High price tag of communication – The amount of energy sensor nodes use for communication is much higher than the amount they use for sensing and computation. This means that techniques such as data compression and aggregation are very important in sensor networks, and special routing techniques are necessary to avoid placing too much of a routing burden on any single node in the network.
- Ad hoc nature of sensor networks – Eventually, some nodes will run out of energy and fall out of the network. The network must be made resilient to such events, through bypassing dead nodes in the routing process. Similarly, if more nodes are added to replenish the sensor network, they must be seamlessly integrated without the need for operator intervention. These considerations firmly place sensor networks in the wider context of ad hoc computer networks.

1.2 Applications of Sensor Networks

Sensor networks began their life as a DARPA project, which materialized in the form of IUSS (Integrated Undersea Surveillance System) later deployed by US Navy in submarine warfare (detection of hostile submarines) [2]. Today, sensor networks have found applications in many diverse fields such as

- Military applications – terrain scanning, imaging, surveillance
- Medicine – remote monitoring of patients, especially the elderly; skin implants for early detection of various illnesses and measurement of blood parameters, wearable computing, swallowable capsules for video imaging of the inside of patient's body

- Automobile traffic – sensors in cars and traffic infrastructure, for congestion monitoring and prevention of road accidents
- Home automation – temperature and humidity measurement, air conditioning automatic control, alarm systems, etc.
- Agriculture – humidity and temperature measurement, automatic control over water sprinklers etc., tracking of cattle movement [3]
- Environmental issues – monitoring for hazardous gases, early earthquake and fire detection

1.3 Issues in Sensor Networks

In this section we give a short overview of some of the open issues in wireless sensor networks, in order to better illustrate the kind of challenges sensor networks researchers are facing.

1.3.1 Energy Conservation

By far the largest issue in sensor networks is battery conservation and careful use of available energy. Some solutions in computer operation which are suitable for desktop computers (with practically unlimited amount of energy at disposal) are proving to be quite inefficient in sensor nodes. One such example is the busy-wait loops and interrupt polling and masking.

Methods for energy conservation in sensor networks can be broadly divided into two groups: passive and active methods.

Passive methods include using sophisticated energy sources to supplant the batteries and placement of sensors into energy-efficient topologies. The alternate energy sources include wind turbines (to allow sensor nodes deployed in the outdoors to run partly using wind power), solar panels, micro-fuel cells, ultracapacitors (special capacitors which hold electrical charge in the dielectric plus additional ionic charge in the double electrical layer, thus increasing energy density by the order of magnitude). In wearable computing use cases, the energy of the movement of the human body can be used as well [4].

Deployment of sensors into energy-efficient topologies benefits from a simple principle: since energy required to bridge a distance d is proportional to the square of that distance, $K * d^2$, then if an intermediate node could be found on half of that distance, $d/2$, the total sum of the energy required to bridge the half-distance twice is $(K * d^2)/2$, which is half of the original energy cost. In energy-efficient topologies, the maximum distance between any two nodes that need to communicate is always under some threshold which is a parameter of the topology (Fig. 1.4).

Active methods to conserve energy include using specialized operating systems (one of which is described in Chapter 4, "Software Aspects of Wireless Sensor Networks"), watchdog timers, which are able to reset the sensor node if the software

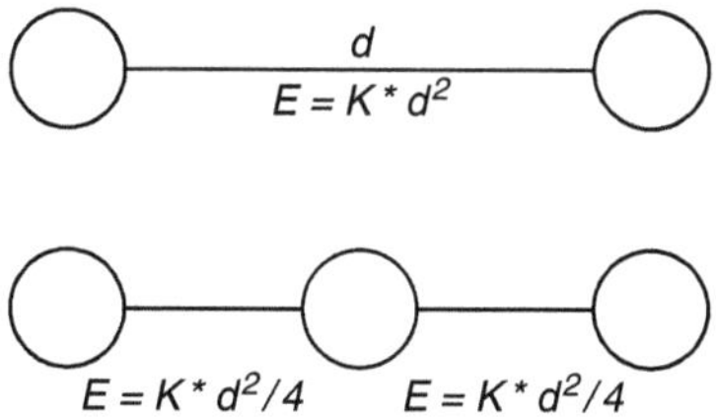

Fig. 1.4 Energy conservation through deployment into energy efficient topologies

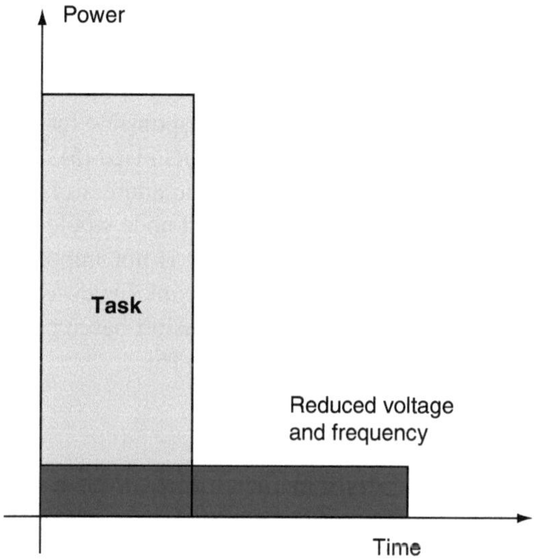

Fig. 1.5 Energy conservation through variable voltage processing

gets stuck in an infinite loop, needlessly expending energy; using sleep states; and finally, variable voltage processing, which exploits the fact that with decreased processing frequency (which allows processor voltage to be slightly dropped), some tasks may take longer to finish, but the power/time product might be smaller thus yielding smaller overall energy consumption (Fig. 1.5) [4].

1.3.2 Routing

Due to energy considerations, the primary criteria for selection of the best route in computer networks – "shortest path" – cannot be directly applied to sensor networks. The reason for this is that in such schemes, the same nodes will always be used to relay traffic to the gateway using the "shortest path," and will eventually run out of energy.

The applied solution is to combine the shortest path metric with a "remaining battery life" metric, i.e., to introduce some sort of reluctance in sensor nodes which is inversely proportional to remaining battery life. In that way, efficiency of routing is partly sacrificed, but the burden of routing is spread more evenly across the network, prolonging its lifetime.

Another important aspect of routing in sensor networks is the issue of addressing. Traditional networks use data-centric addressing – the basis for the addressing scheme is some kind of datum, which in IP networks is the 4-byte number known as an IP address. Most of the sensor network architectures use a similar approach; for example, in TinyOS each node is assigned a 1-byte number address, with the gateway node being assigned the address 0. However, in some use-cases, due to unstable nature of sensor networks (and ad hoc networks in general) it is useful to have geographical addressing: A message directed to a certain area would always reach a node positioned in that area, even if node previously responsible for that area had already used up its batteries and another node had taken its place [5]. This would not work with the data-centric approach, as the data-centric address of the new node would not be the same; the message addressed to the old node would be lost.

Finally, in some sensor networks addressing is not important at all; the only important thing is to diffuse collected data to the sink (gateway node) so they could be processed remotely. For information on a diffusion-based protocol, an interested reader is directed to the Rumor Routing protocol [6].

1.3.3 Positioning and Localization

In many applications of sensor networks, it is important to record the actual location where the measurement was made. For example, if there is a sensor network monitoring the gradient of the temperature, each temperature measurement would have to be accompanied by the information on where it was recorded. For this to happen, each sensor node would have to have knowledge of its location. In accepted terminology, **positioning** the sensor node means providing it with **absolute** coordinates (such as longitude and latitude); **localizing** the sensor node means providing it with information on its position **relative** to other sensor nodes.

An obvious choice is to provide all sensor nodes with a GPS device. While very straightforward, this approach has significant drawbacks: GPS devices cost money (a $50 GPS module deployed on 100 sensor nodes would increase the network cost by $5,000), plus they significantly increase the rate of energy consumption, shortening the battery life considerably.

Fortunately, it is not necessary to provide all nodes with a GPS if we want to have full positioning information. In a 2-D case, if at least three devices are provided with GPS information (either through a GPS device or if GPS positions are input manually) all other devices can gradually establish positioning information through a simple algorithm:

- Record distances d_1 and d_2 to two nodes which have positioning information.
- Draw circles with diameters d_1 and d_2 around these two nodes.
- These circles will intersect in two points, n and n'.
- Check mutual visibility between the new node and the third already positioned node, in order to discount either of these two points.
- Position the new node in the remaining of these points.

For the 3-D case, a similar algorithm is used: Spheres are drawn around three already positioned nodes; these three spheres can intersect in up to two points. However, if all sensor nodes are physically on the earth's surface, then one of these points can usually be discounted on the basis that it is either above or beneath the surface, so the final step can be omitted.

If no nodes have information on their absolute position, the positioning approach is impossible; however, if physical distances between any two nodes are known (one way to measure physical distance is through measurement of radio attenuation) then localization is possible. Localization means establishing an internal coordinate system, with origin in one of the nodes (which would then have coordinates 0, 0, 0) and assigning coordinates to all other nodes, using the same system. For an example localization algorithm, an interested reader is directed to [7].

1.3.4 Time Synchronization

In addition to position information, a sensor measurement usually needs to be accompanied by exact time in which the measurement had been made. At first glance, this does not seem to be much of an issue. However, in applications where precision is paramount, it is very important to keep clocks on all sensor nodes synchronized, as a deviation of even a few milliseconds could compromise the validity of the measurement.

In traditional local area networks, the solution is to use the clock on the server and to keep clocks on workstations synchronized to that clock, all the time. This is made possible through reliable, wired communication with predictable delays (a workstation measures round-trip time, asks for the time from the central server, and uses measured RTT to compensate for the delay). However, in sensor networks, this approach is not suitable, because synchronizing clocks regularly means spending a lot of energy to relay clock synchronization messages. Furthermore, communication delays are not so predictable.

To this day, a "perfect" solution for the time synchronization issue in sensor networks has not been found. Some of the ideas that are used for time synchronization include:

- Explicit synchronization – Clocks are not kept synchronized all the time; instead, in order to save on the communication overhead, each sensor node keeps its own timescale, and conversion between different time scales is done "on demand" (for example, at each hop of data routing).

- Peer-to-peer synchronization – Clocks are only kept synchronized between neighboring sensor nodes. The rationale for this is that communication between neighboring nodes involves only those nodes themselves (as opposed to a number of intermediate nodes which route sync messages in central synchronization).

For more information on time synchronization in sensor networks, an interested reader is kindly asked to consult [8].

1.3.5 Security

Sensor networks typically consist of a very large number of nodes, so supervision of each and every node is practically impossible. As a result, sensor networks are highly susceptible to logical and physical attacks and communication interception. For instance, a node could be seized, reprogrammed, and returned to the network; or using reverse engineering, nodes could be built with an ability to "trick" the network into treating them as authentic. Various forms of sensor network abuse are possible: from eavesdropping on sensitive information, imputing falsified readings, to Distrib uted Denial of Service (DDoS) attacks.

Since it is unfeasible to protect every single node, other approaches are used [9]:

- Node-to-node authentication (nodes have to prove their identity to each other)
- Node revocation: when an intruding node is discovered, it is forbidden to access the network any further
- Resilient protocols in the network – designed to continue working properly even with a number of compromised nodes

As in traditional networks, privacy of sensed data is maintained by encryption. However, the conventional approach – the use of very large keys – is unsuitable for sensor networks, because of limited memory capacity. Instead, two commonly used approaches are **hop-to-hop encryption** and **multipath routing**.

In hop-to-hop encryption, a message is encrypted using different short keys in each node along the path from source to destination. One drawback of such a scheme is that the takeover of any of the nodes on the path, while not exposing the contents of the message, prevents any further communication along the path, as the chain of encryption will be broken.

In multipath routing, messages are broken into several "chunks" before sending. These chunks move along different routes and are not reassembled until they reach the destination. In this way, message can be intercepted only if at least one node on each path is compromised.

The final threat to the security in sensor networks are DDoS (Distributed Denial of Service) attacks. Through these attacks, attackers can deliberately drain the batteries of sensor nodes. Protection against these attacks can be physical and logical. Physical means of protection is primarily spread spectrum communication (direct sequence or frequency hopping); logical means are constant checking of incoming messages for authencity and discarding messages with invalid authencity information.

One danger with this approach is that the very protection from DdoS attacks can drain the battery, thus accomplishing the goal of the attacks themselves as constant checks for message authencity extract a high energy toll from the batteries.

1.4 Conclusion

Sensor networks have come a long way from obscure military beginnings into a widely discussed important research subject, and in the twenty-first century first everyday applications are slowly emerging. Since sensor nodes are in concept small computers with high interconnection ability, a lot of solutions developed for traditional computers (such as routing, time synchronization, and security) could be reused, but with substantial adaptations due to significantly smaller processing power and memory sizes smaller by several orders of magnitude. On the other hand, some other concepts, such as positioning and localization, were never necessary in traditional computers and thus had to be developed from scratch.

References

1. Dr. Thomas Hanselmann (2007) "Sensor Networks" http://pi4.informatik.uni-mannheim.de/~haensel/sn_book.pdf
2. Integrated Undersea Surveillance System (IUSS) – http://www.fas.org/irp/program/collect/iuss.htm
3. Camalie Vineyards, www.camalie.com
4. Ilyas M, Mahgoub I (eds) (2005) Handbook of sensor networks: compact wireless and wired sensing systems. CRC Press, Boca Raton, FL
5. A Survey of Geocast Routing Protocols, http://www.comsoc.org/livepubs/surveys/public/2004/apr/maihofer.html
6. Braginsky D, Estrin D (2002) Rumor routing algorithm for sensor networks Proceedings of the 1st ACM international workshop on Wireless sensor networks and applications, Atlanta, Georgia, USA
7. Capkun S, Hamdi M, Hubaux J (2001) GPS-free positioning in mobile ad hoc networks. IEEE, Washington, DC
8. Elson J (2003) Time synchronization in wireless sensor networks, UCLA
9. Chan H, Perrig A (2003) Security and privacy in sensor networks, IEEE Computer, October 2003, 36:103–105

Part I
Basic Issues

Chapter 2
Sensors

Goran Rakočević

2.1 Introduction

A sensor is an electronic device used to detect or measure a physical quantity and convert it into an electronic signal. In other words, sensors are devices that translate aspects of physical reality into representations understandable and processable by computers.

In a wireless sensor network, sensors play an important part, as sensing is one of its central roles. Technology behind sensors, however, is not of major interest when considering sensor networks, with the emphasis being more on communication, network management, and data manipulation. Most sensors used in WSN systems have been developed independently of WSN technology, and these two fields continue to develop somewhat independently. Nevertheless, any in-depth discussion of wireless sensor networks, especially when aimed toward providing the reader with a holistic picture of current capabilities and limitations of wireless sensor networks, must include sensors.

In this chapter, first we will look at some important issues regarding sensors and introduce some of the terminology used later in the chapter. Then we move on to examine basic types of sensors, categorized by the function they perform. Characteristics of currently available sensors are given, along with a brief overview of their operating principals. The goal is not to go into a detailed analysis of sensor technology, but to give the reader a basic notion of how a certain sensor operates, under which conditions, and with which limitations. This insight should prove handy when designing a new WSN application. For a more detailed study of sensors and sensing technologies, readers are referred to [1].

Finally, as this book puts much emphasis on utilization of WSN technology in personal and public health, we examine some of the more complex sensing devices used within this field.

G. Rakočević (✉)
School of Electrical Engineering,
University of Belgrade
e-mail: grakocevic@gmail.com

L. Gavrilovska et al. (eds.), *Application and Multidisciplinary Aspects of Wireless Sensor Networks*, Computer Communications and Networks,
DOI 10.1007/978-1-84996-510-1_2, © Springer-Verlag London Limited 2011

2.2 Sensor Characteristics

In this chapter we introduce some of the important characters of sensors. Understanding of these is important when choosing the right sensors for your application.

2.2.1 *Transfer Function*

Transfer function is a mathematical representation of the relation between the input and output of a system. In terms of sensors it represents the relationship between the physical signal on the input, and the electrical output signal.

2.2.2 *Hysteresis*

A sensor should be capable of following changes in the input parameter, regardless of what direction the parameter changes in (whether there is an increase or a decrease in value). Some sensors do not return the same value for both directions of change. Hysteresis is the measure of this property.

2.2.3 *Linearity*

Linearity of a sensor shows how much the measured curve deviates from the ideal transfer function. There are several ways to interpret and represent linearity, the most commonly used being:

- End point linearity
- Best-fit straight line (BFSL)
- Least squares BFSL linearity

On the Input against Output graph a straight line is drawn from the zero point to the full-scale output point. The point on the actual measured curve that deviates most from this line is used to define the linearity of the sensor. This is usually quoted as a percentage of the full-scale output of the sensor.

2.2.3.1 Best-Fit Straight Line (BFSL)

BFSL linearity is a measure of the most accurate overall results that can be obtained for a given sensor. Instead of using a line that connects the full-scale point to the end point, a different line is chosen so that the maximal deviation of the actual measured curve is minimal. Note that this is simply a means of interpretation of the results and relies entirely on the system, including the measuring equipment, being set up to take advantage of this technique.

2.2.3.2 Least Squares BFSL Linearity

The method of least squares assumes that the best-fit curve of a given type is the curve that has the minimal sum of the deviations squared (least square error) from a given set of data.

When defining least squares BFSL linearity, the slope of the best-fit line is defined using the measured results in a number of calibration points with the equation:

$$\frac{\Sigma(\text{actual sensor output at each data point}) * (\text{actual sensor output at each data point})}{\Sigma(\text{actual sensor output at each data point})^2}$$

Having mathematically determined the slope of the best-fit straight line it is then possible to determine the maximum deviation of any point from this line.

2.2.4 Sensitivity

Sensitivity of a sensor is the ratio between a small change of the input and the resulting change in the output signal. Mathematically speaking, it is defined as the slope of the output characteristic curve. Sensitivity error is a departure from the ideal slope of the characteristic curve.

2.2.5 Accuracy

Accuracy represents the largest expected error between the ideal output signal and the actual output signal. Sometimes it is presented as a percentage of the maximum output signal.

2.2.6 Dynamic Range

Dynamic range (or Span) is the range of the input signal that can be accurately converted into the electrical output. Outside of the dynamic range sensor produces either a predefined value, or, more commonly, undefined and inconsistent.

2.2.7 Noise

All sensors produce noise in addition to output signal. For applications that require high precision, sensing amount of noise introduced by a sensor can be of at most importance.

2.2.8 *Resolution*

The resolution of a sensor is the minimum fluctuation of the input signal that can be detected.

2.2.9 *Bandwidth*

After any change in the input parameter, a certain amount of time is required before the resulting change in the output parameter can be detected and measured. This time is called response time. Some sensors also many sensors have decay times, which would represent the time after a step change in physical signal for the sensor output to decay to its original value. Values reciprocal to the response time and the decay time are called lower and upper cutoff frequencies. The bandwidth of a sensor is the frequency range between these two frequencies. A sensor with high bandwidth can change its output to follow fast (high-frequency) variations in the input. For measures that change slowly, such as the temperature of a large liquid tank, bandwidth may not be important. For other applications such as scanning shape or vibration sensors, bandwidth may be the most important figure of merit.

2.3 Classifications

There are two basic ways to categorize sensors. The first is based on the principal by which they function, and the second is based on the function the sensor performs. Most sensors act like passive devices (i.e., capacitors or resistors). These sensors require external circuitry for biasing and amplification of the output signal.

Resistive sensors are devices whose resistance changes with the value of input signal being measured. These sensors can be used in a simple voltage-divider configuration (Fig. 2.1). For more precise measurements a variety of configurations can be used (e.g., the Whetstone bridge circuit).

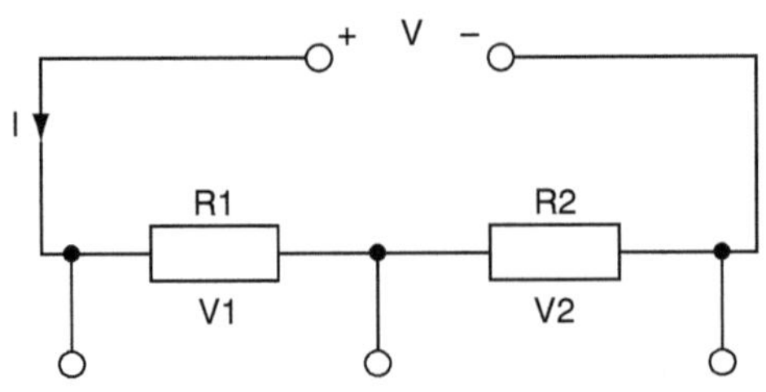

Fig. 2.1 Voltage divider. Legend: R1 – Resistive sensor, R2 – Reference resistor, V1 – Voltage on the resistive sensor, V2 – Voltage on the reference resistor, V – reference voltage

Capacitive sensors produce a change in capacitance proportionate to the value of the measured input signal. Detection of this change is done quite similarly as with the resistive sensors, only in this case the impedance of the capacitor is observed, which means that an AC bias must be provided. Inductance-based sensors can be observed in much the same way.

As opposed to these sensors some sensors produce their own bias voltage, and can directly be connected to an AD converter or an amplifier, if amplification is required.

Perhaps the more logical way to classify sensors is with regard to the physical property they measure. The most common categories include [2]:

- Mechanical
- Thermal
- Electrical
- Magnetic
- Radiant
- Chemical and biochemical

2.4 Mechanical Sensors

Mechanical sensors detect mechanical properties and actions. These include (among other things) pressure, velocity, vibration sensors, and accelerometers.

2.4.1 Pressure Sensors

Pressure is one of the most important physical properties and, thus, pressure micro-sensors were the first micro-sensors developed and used by the industry. A wide variety of applications calls for a wide variety of pressure sensors, but most belong in one of three major categories.

Piezorezistive pressure sensors have a piezoresistor integrated in a membrane. Pressure is applied to the membrane, causing it to deform. This in turn, causes a change in resistance, proportionate to the applied force.

In capacitive pressure sensors (whether membrane or comb-based) pressure is applied on the sensor surface, causing a membrane to deflect and the capacitance to change. These sensors generally have greater sensitivity and linearity, while exhibiting very little or no hysteresis. However, these sensors also have higher production costs when compared to piezoresistive pressure sensors.

2.4.2 Optical Pressure Sensors

Optical pressure sensors operate on the principal of the Mach–Zehnder interferometer. Laser light is brought into the sensor via an optical fiber. This light is split into two beams. One of the two beams crosses through one of the beams, which is deformed by the pressure. This deformation changes the light's properties. The two beams are combined and brought to a photodiode. Different propagation speeds create a phase shift between these beams which is detected at the diode.

2.4.3 Position and Motion Sensors

Position sensors play an important role in a wide variety of applications. Numerous ways of detecting position are available, ranging from simple contact sensors to more complex contact-free ones. Position measurement can either be relative (displacement sensors) or absolute, linear or angular.

All types of position sensors have their own advantages and drawbacks; thus, it is important to choose the right type of the sensor for the given application.

2.4.4 Resistive Position Sensors

Resistive position sensors are better known as potentiometers. In various forms, these sensors have found applications from volume adjustment knobs and sliders on radios to industrial machine slide sensing. A potentiometer is simply a resistor whose resistance changes based on the position of a movable part.

2.4.5 Limit Switches

Limit switches are the simplest of position sensors. They are electromechanical contact devices. A limit switch consists of a number of switches. When the monitored object comes into contact with one of the switches, this switch is activated.

2.4.6 Magnetic Position Sensors

Magnetic Position Sensors are noncontact position sensors that are magnetic-fields-generated or affected by target objects.

The magnetic field is a vector quantity that has both magnitude and direction. Scalar magnetometers measure only the total magnitude. Others measure the magnitude of the component of magnetization along their sensitive axis (unidirectional sensors). Vector magnetic sensors have two or three bidirectional sensors.

2.4.7 *Hall Effect Sensors*

A Hall Effect sensor is a transducer that varies its output voltage in response to changes in magnetic field.

In its simplest form, the sensor operates as an analogue transducer, directly returning a voltage. They contain a Hall element constructed from a thin sheet of conductive material with output connections perpendicular to the direction of current flow. With a known magnetic field, its distance from the Hall plate can be determined. Using groups of sensors, the relative position of the magnet can be deduced.

Electricity carried through a conductor will produce a magnetic field that varies with current, and a Hall sensor can be used to measure the current without interrupting the circuit. Typically, the sensor is integrated with a wound core or permanent magnet that surrounds the conductor to be measured.

2.4.8 *Magnetoresistive Sensors*

Magnetoresistance is the property of a material to change the value of its electrical resistance when an external magnetic field is applied to it.

MR sensors measure both linear and angular position and displacement in the Earth's magnetic field (below 1 gauss). They are an excellent solution for locating objects in motion. By affixing a magnet or sensor element to an angular or linear moving object with its complementary sensor or magnet stationary, the relative direction of the resulting magnetic field can be quantified electronically.

2.4.9 *Ultrasonic Position Sensors*

Ultrasonic sensors work on a principle similar to radar or sonar which evaluates attributes of a target by interpreting the echoes from radio or sound waves, respectively. Ultrasonic sensors generate high-frequency sound waves and evaluate the echo which is received back by the sensor. Sensors calculate the time interval between sending the signal and receiving the echo to determine the distance to an object.

2.4.10 *Accelerometers*

Accelerometers are sensors that measure acceleration they are subjected to. Most are based on resistive or capcitative and piezoelectric methods.

2.4.11 *Resistive and Capacitive Accelerometers*

With these micro-sensors an elastic cantilever with an attached mass is usually used. When the sensor is subjected to acceleration, a force proportionate to this acceleration deforms the cantilever. With piezoresistive sensors a piezoresistor is integrated into the cantilever, whose deformation causes a change in its resistance. With capacitive sensors the cantilever acts as one electrode, with an electrode strip acting as the other. As the cantilever is deformed it is brought closer to the electrode strip, which in turn affects the capacitance between the two electrodes.

Resistive and capacitive accelerometers can be used to measure constant acceleration, such as that of Earth's gravity. They are generally used for measuring low-frequency vibrations.

2.4.12 *Piezoelectric Accelerometers*

Piezoelectric accelerometers are based on the piezoelectric effect. This means that an electric charge is created when the sensing material is squeezed or strained. Several methods of straining of the material can be used, three of the basic being: compression, flexural, and shear, with the shear being the most common one. These accelerometers are generally durable, protected from contamination, and impervious to extraneous noise influences.

2.5 Temperature Sensors

Temperature is the most widely sensed of all parameters. Temperature sensors detect a change in a physical parameter (resistance or output voltage) that corresponds to a temperature change. Three basic types of temperature sensors are electromechanical, electronic, and thermo-resistive [3].

2.5.1 *Electromechanical Temperature Sensors*

These sensors are based on expanding or contracting properties of materials when subjected to a temperature change. Bi-metal thermostats are created by bonding two metals into a single strip of material. Different expansion rates of the metals create electromechanical motion when the material is subjected to a temperature change. In capillary thermostats the capillary motion of expanding or contracting fluid is used to make or break a number of electrical contacts.

2.5.2 *Electronic Temperature Sensors*

Thermocouples: Thermocouples are based on the Seed-back effect. When a pair of dissimilar metals is joined at one end, and there is a temperature difference between the joined end and the open end, thermal electromagnetic force is generated. This will create a flow of current through the wires that is proportionate to the temperature difference. The open ends must be kept at a constant reference temperature. Several standard types of thermocouples are use.

2.5.3 *Silicon Sensors*

Silicon sensors make use of the bulk electrical resistance properties of semiconductor materials, rather than the junction of two differently doped areas. Especially at low temperatures, silicon sensors provide a nearly linear increase in resistance versus temperature or a positive temperature coefficient.

2.5.4 *Resistive Temperature Sensors*

Resistive temperature sensors are devices whose resistance changes with the temperature.
Thermistors: A thermistor is a type of resistor with resistance varying according to its temperature. They typically consist of a combination of two or three metal oxides that are sintered in a ceramic base material.

Thermistors can be classified into two types: positive temperature coefficient (PTC) and negative temperature coefficient (NTC). PTC devices exhibit an increase in resistance as temperature rises, while NTC devices exhibit a decrease in resistance when temperature increases.

The main disadvantage of the thermistor is its strong nonlinearity. Cheap thermistors have large spread of parameters ("tolerance") and calibration is usually necessary.

2.5.5 *Resistive Temperature Detectors (RTDs)*

Unlike thermistors that use a combination of metal oxides and ceramics, resistive temperature detectors are made from pure metal (copper, nickel, or platinum are usually used). RTDs are useful over larger temperature ranges, while thermistors typically achieve a higher precision within a limited temperature range.

As a RTD is a resistance device, it needs measuring current to generate a useful signal. Because this current heats the element above the ambient temperature ($P = I^2R$), errors can occur, unless the extra heat is dispersed. This forces us to choose a small-sized resistance device with a quick response or a larger resistance device and better heat release.

A second solution is to keep the measuring current low (usually between 1 and 5 mA).

2.6 Humidity Sensors

Humidity is the amount of water vapor in the given substance (usually a gas). It is an important parameter in a variety of fields, including room air humidity in patient monitoring and exhibit perseveration in museums, meteorological observations, soil humidity in agriculture, and process control in industrial applications.

Humidity can be measured as the absolute humidity (ratio of water vapor to the volume of substance), relative (compared to the saturated moisture level) or dew point (temperature and pressure at which the observed gas starts to turn into liquid). Most common humidity sensors are based on capacitive, resistive, and thermal conductivity measurement techniques.

2.6.1 Capacitive RH Sensors

In a capacitive RH sensor, change in dielectric constant is almost directly proportional to relative humidity in the environment. Relative humidity sensors have three-layer capacitance construction and consist of thermoset polymer, platinum electrodes, and a silicon chip with integrated voltage output signal conditioning.

These sensors have low temperature coefficient, and response times that range from 30 to 60 s. They offer near-linear voltage outputs, wide RH ranges and condensation tolerance, and are stable over long-term use. However, the capacitive effect of the cable connecting the sensor to the signal conditioning circuitry is large compared to the small capacitance changes of the sensor. This limits the distance from sensing element to signal conditioning circuitry.

2.6.2 Resistive Humidity Sensors

Resistive humidity sensors measure the resistance change in a medium such as a conductive polymer or a salt. Resistance usually has an inverse exponential relationship to humidity. Response times of these sensors are 10–30 s.

Resistive humidity sensors are small size, low cost, and are usable from remote locations.

2.7 Chemical Sensors

Chemical sensors detect the presence or concentration of particular chemical elements or compounds in a given sample. A chemical sensor usually consists of a chemically sensitive film or a membrane and a transducer.

A chemical process occurring in or on a chemically sensitive film or membrane causes a signal to be generated at the transducer. Examples of mechanisms commonly employed include host–guest binding, catalytic reactions, or a redox process.

Chemical sensors have a vast variety of applications ranging form medical diagnostics and nutritional sciences, through security to automotive industry (Fig. 2.2).

Based on the principal of operation, several types of chemical sensors can be identified.

2.7.1 Interdigital Transducer Sensors

Interdigital transducers using capitative measurement are often used in chemical sensors. Sensitive layer is used as the dielectric between two electrodes. The dielectric properties of the sensitive layer are changed when it interacts with certain substances, affecting the capacitance between the two electrodes.

2.7.2 Conductivity Sensors

In these sensors the sensitive layer is used as a conductor of electricity. Interactions with certain chemicals (e.g., absorption of gasses) modify the conductivity of this layer. There are two types of sensing layers: Metal Oxide and Conducting Polymers.

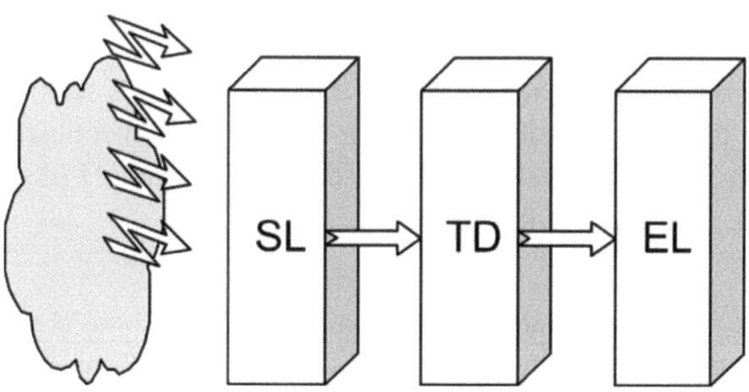

Fig. 2.2 Structure of a chemical sensor. Legend: CS – Chemical substance, SL – Sensitive layer, TD – Transducer, EL – Electronics. Explanation: Chemical substance reacts with the chemical layer. Reaction causes a signal to be generated at the transducer. The signal is then processed by electronics and converted into a format suitable for further processing

Metal Oxide sensitive layers are typically made of SnO_2 doped with Pt or Pd. These sensors can operate at high temperatures (300–5,000°C), which makes them especially suitable for combustion gases.

Conductive Polymer sensitive layers are usually based on pyrrole, aniline, or thiophene. These sensors operate best at room temperatures. Compared to Metal Oxide sensors these sensors have lower power consumption, and faster response and recovery times. However, they are have lower sensitivity and are sensitive to humidity.

2.7.3 *Optical Chemical Sensors*

In optical sensors, an optical waveguide is used as the sensitive layer. Chemical reactions between the waveguide and the target chemical substance cause a change in the optical properties of the waveguide (e.g., the index of reflection). As a result the amount (or the wavelength) of the light striking the sensor on the end of the waveguide varies.

These sensors are highly sensitive, can handle small quantities, are inexpensive, and easy to sterilize.

Majority (about 60%) of chemical sensors are gas sensors. Most commonly used chemical sensors include O_2, pH, CO, CO_2, NO_X, Methane, etc. Table 2.1 gives an overview of the range of characteristics for some of these sensors available on the market.

2.7.4 *Ion-Sensitive FET Sensor*

An ion-sensitive field effect transistor (ISFET) is an ion-sensitive field effect transistor used to measure ion concentrations in solution; when the ion concentration (such as pH) changes, the current through the transistor will change accordingly. Here, the solution is used as the gate electrode. A voltage between substrate and oxide surfaces arises due to an ions sheath.

An ISFET's source and drain are constructed as for a MOSFET. The gate electrode is separated from the channel by a barrier which is sensitive to hydrogen ions and a gap to allow the substance under test to come in contact with the sensitive barrier. An ISFET's threshold voltage depends on the pH of the substance in contact with its ion-sensitive barrier.

Table 2.1 Overview of important characteristics for some chemical sensors

Sensor type	Response time	Range	Accuracy	Temperature range
Oxygen	4 s	0–150 mm-Hg		
CO	20 s	0–5,000 ppvmol	±5%	
NO_x	<40 s	0–100 ppm		−20°C to 50°C
H_2S	<60 s	0–200 ppm	±2%	

The surface hydrolization of OH groups of the gate materials varies in aqueous solutions due to pH value. Typical gate materials are Si_3N_4, Al_2O_3, and Ta_2O_5.

ISFET sensors are used in devices for continuous measurements like those for continuous measurement of pH value and gases in blood (O_2, CO_2).

2.7.5 Piezoelectric Chemical Sensors

Piezoelectric effect is the generation of an electric charge in a crystalline material upon subjecting it to stress. A piezoelectric chemical sensor is a piezoelectric oscillator that responds to changes in the chemical composition of its environment with changes of the resonant frequency, or wave speed [4].

The complex nature of these sensors makes them unsuitable for a brief overview of operating principals, as is suitable for this book. However, as these sensors are undergoing a rapid expansion, readers are encouraged to turn to references for more detailed explanations.

2.8 Biosensors

Detection of presence and concentrations of bacteria, viruses, or molecules and molecular complexes like proteins, enzymes, antibodies, DNA, etc. is essential to a wide range of applications. Traditionally, this has been done through time-consuming chemical analysis methods that require laboratory conditions and employ expensive reagents and equipment. Technological advancements and introduction of micro-sensor technology to this field has led to development of biosensors.

Much like a chemical sensor, biosensor consists of three parts: a sensitive layer, transducer, and electronic circuitry to process the signal from the transducer.

Sensitive layer in a biosensor is a biosensitive biological component, like enzymes, antibodies, cell membrane receptors, tissue slices, etc.

One example of biosensors is imunno-sensors. These sensors use specific antibodies as the sensitive layer. These immobilized antibody molecules bond with antigen molecules in the substance. Concentration of antigens can then be measured using, for example, interferometric method.

Biosensors are under constant and rapid development. Currently, design and production of a biosensor face serious difficulties, most related to immobilization of proteins and long-term stability of proteins.

2.8.1 Radiation Sensors

Ionizing radiation consists of subatomic particles or waves that are energetic enough to detach electrons from atoms or molecules, ionizing them. Exposure to radiation causes microscopic damage to living tissue, resulting in skin burns and radiation sickness at high doses and cancer, tumors, and genetic damage at low doses. Therefore, monitoring radiation levels is imperative in many industrial applications where human interaction with radioactive materials exists, as well as in guarding against intentional or accidental exposure of wider population to radiation. Wireless sensor networks provide ideal infrastructure for these kinds of systems.

2.8.2 Geiger–Müller Counter

Geiger counters are used to detect radiation usually gamma and beta radiation, but certain models can also detect alpha radiation. The sensor is a Geiger–Müller tube, an inert gas-filled tube (usually helium, neon, or argon with halogens added) that briefly conducts electricity when a particle or photon of radiation temporarily makes the gas conductive. The tube amplifies this conduction by a cascade effect and outputs a current pulse.

2.8.3 Quartz Fiber Dosimeter

A quartz fiber dosimeter is a pen-like device that measures the cumulative dose of ionizing radiation received by the device. The device is mainly sensitive to gamma and X-rays, but it also detects beta radiation above 1 MeV. Neutron-sensitive versions have been made.

The quartz fiber dosimeter is a rugged form of a device called a Lauritsen electroscope. It consists of a sealed air-filled cylinder called an ionization chamber. Inside it is a metal electrode strip that is attached to a terminal on the end of the pen for recharging. The other end of the electrode has a delicate gold-plated quartz fiber attached to it, which at rest lies parallel to the electrode.

During recharging, the charger applies a high DC voltage, to the electrode, charging it with electrostatic charge. The quartz fiber, having the same charge, is repelled by the surface of the electrode due to the coulomb force and bends away from the electrode. After charging, the charge remains on the electrode because it is insulated.

When a particle of ionizing radiation passes through the chamber, it collides with molecules of air, knocking electrons off them and creating positively and negatively charged atoms (ions) in the air. The ions of opposite charge are attracted to the electrode and neutralize some of the charge on it.

Since each radiation particle allows a certain amount of charge to leak off the electrode, the position of the fiber at any time represents the cumulative radiation that has passed through the chamber since the last recharge. Recharging restores the charge that was lost and returns the fiber to its original deflected position.

2.8.4 Film Badge Dosimeter

The film badge dosimeter, or film badge, is a dosimeter used for monitoring cumulative exposure to ionizing radiation. The badge consists of two parts: photographic film, and a holder.

The film is sensitive to radiation and, once developed, exposed areas increase in optical density (i.e., blacken) in response to incident radiation. One badge may contain several films of different sensitivities or, more usually, a single film with multiple emulsion coatings. This allows for separate measurement of neutron, beta, and gamma exposure, and estimation of energy spectra. The holder may contain a number of filters that attenuate certain types of radiation, such that only the target radiation is monitored. To monitor gamma rays or X-rays, the filters are metal, usually tin or lead. To monitor beta particle emission, the filters use various densities of plastic.

2.8.5 Thermoluminescent Dosimeter

A thermoluminescent dosimeter, or TLD, is a type of radiation dosimeter. A TLD measures ionizing radiation exposure by measuring the amount of visible light emitted from a crystal in the detector when the crystal is heated. The amount of light emitted is dependent upon the radiation exposure.

A TLD is a phosphor, such as lithium fluoride (LiF) or calcium fluoride (CaF), in a solid crystal structure. When a TLD is exposed to ionizing radiation at ambient temperatures, the radiation interacts with the phosphor crystal and deposits all or part of the incident energy in that material. Some of the atoms in the material that absorb that energy become ionized, producing free electrons and areas lacking one or more electrons, called holes. Imperfections in the crystal lattice structure act as sites where free electrons can become trapped and locked into place.

Heating the crystal causes the crystal lattice to vibrate, releasing the trapped electrons in the process. Released electrons return to the original ground state, releasing the captured energy from ionization as light, hence the name thermoluminescent. Released light is counted using photomultiplier tubes and the number of photons counted is proportional to the quantity of radiation striking the phosphor.

Instead of reading the optical density (blackness) of a film, as is done with film badges, the amount of light released versus the heating of the individual pieces of thermoluminescent material is measured.

2.9 Medical Sensing Devices

This book puts a special emphasis on medical applications of wireless sensor networks. Therefore, a section of this chapter is devoted to devices used for monitoring human vital parameters.

2.9.1 *Pulse Oximetry*

A pulse oximeter is a medical device that indirectly measures the oxygen level in a patient's blood and changes in blood volume in the skin, producing a photoplethysmograph. As noninvasive measurement instruments they are particularly convenient for wide use.

Typically a pulseoximeter has a pair of small light-emitting diodes (LEDs) facing a photodiode through a translucent part of the patient's body, usually a fingertip or an earlobe. One of the LEDs is red, with wavelength of 660 nm, and the other is infrared, 905, 910, or 940 nm. Absorption at these wavelengths differs significantly between oxyhemoglobin and its deoxygenated form; therefore, from the ratio of the absorption of the red and infrared light, the oxy/deoxyhemoglobin ratio can be calculated. The absorbance of oxyhemoglobin and deoxyhemoglobin is the same (isosbestic point) for the wavelengths of 590 and 805 nm.

The monitored signal bounces in time with the heart beat because the arterial blood vessels expand and contract with each heartbeat. By examining only the varying part of the absorption spectrum, a monitor can ignore other tissues or nail polish and discern only the absorption caused by arterial blood. Thus, detecting a pulse is essential to the operation of a pulse oximeter and it will not function if there is none (Fig. 2.3).

A pulse oximeter is useful in any setting where a patient's oxygenation is unstable, including intensive care, operating, recovery, emergency and hospital ward settings, pilots in unpressurized aircraft, for assessment of any patient's oxygenation, and determining the effectiveness of or need for supplemental oxygen. Assessing a patient's need for oxygen is the most essential element to life; no human life thrives in the absence of oxygen (cellular or gross). Although a pulse oximeter is used to monitor oxygenation, it cannot determine the metabolism of oxygen, or the amount of oxygen being used by a patient. For this purpose, it is necessary to also measure carbon dioxide (CO_2) levels.

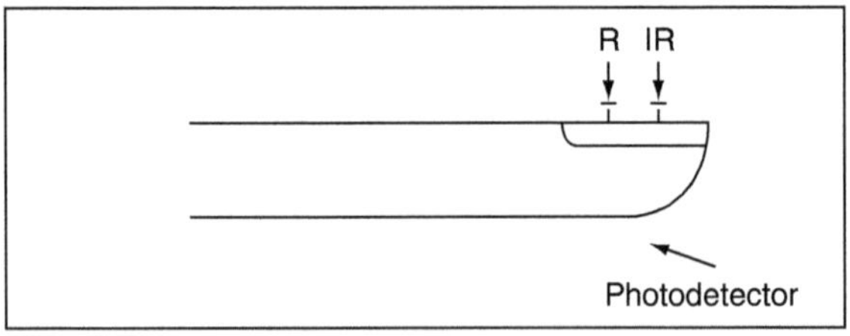

Fig. 2.3 Operating principle of a pulseoximeter. Explanation: Red and infrared light emmitors are positioned above the fingernail. A photodetector is positioned below the finger. Absorption of red and infrared light differs significantly between oxyhemoglobin and its deoxygenated form; therefore, from the ratio of the absorption of the red and infrared light, the oxy/deoxyhemoglobin ratio can be calculated

Because of their simplicity and speed, pulse oximeters are of critical importance in emergency medicine and are also very useful for patients with respiratory or cardiac problems, especially COPD, or for diagnosis of some sleep disorders such as apnea and hypopnea. Portable, battery-operated pulse oximeters are useful for pilots operating in a non-pressurized aircraft above 10,000 ft (12,500 ft in the US), where supplemental oxygen is required.

A pulse oximeter does, however, have a number of drawbacks, limiting their use, or calling for caution when oximetry results are used and interpreted.

First, it should be noted that oximetry is not a complete measure of respiratory sufficiency. A patient suffering from hypoventilation (poor gas exchange in the lungs) given 100% oxygen can have excellent blood oxygen levels while still suffering from respiratory acidosis due to excessive carbon dioxide.

It is also not a complete measure of circulatory sufficiency. If there is insufficient blood flow or insufficient hemoglobin in the blood (anemia), tissues can suffer hypoxia despite high oxygen saturation in the blood that does arrive.

Furthermore, a high level of methemoglobin will tend to cause a pulse oximeter to read closer to 85% regardless of the true level of oxygen saturation. It also should be noted that the inability of two-wavelength saturation level measurement devices to distinguish carboxyhemoglobin due to carbon monoxide inhalation from oxyhemoglobin must be taken into account. To counter this problem, CO-oximeters have recently been developed. These devices use additional wavelengths to distinguish CO from O_2.

2.9.2 *Continuous Glucose Monitors*

A continuous blood glucose monitor (CGM) determines blood glucose levels on a continuous basis (every few minutes). These devices include a disposable glucose sensor placed just under the skin and worn for a few days until replacement, a link from the sensor to a non-implanted transmitter which communicates to a radio receiver and an electronic receiver worn like a pager (or insulin pump) that displays blood glucose levels on a practically continuous manner.

The sensor can be implanted in the subcutaneous tissue using a specialized tool designed to minimize tissue damage. The tip of the sensor is made of a membrane selectively permeable to glucose. Once the glucose passes through the membrane, it is oxidized by the enzyme glucose oxidase. Reduced glucose oxidase can then be oxidized by reacting with molecular oxygen, forming hydrogen peroxide as a by-product. At the electrode surface, hydrogen peroxide is oxidized into water, generating a current which can be measured and correlated to the glucose concentration outside the membrane (see above right).

This type of device requires at least four finger sticks per day for calibration. Continuous glucose monitors measure glucose levels in the interstitial fluid, not directly the blood. This causes a temporal lag behind actual blood glucose values of about 5 min.

2.10 Electrocardiograph

An electrocardiograph is a device for recording of the electrical activity of the heart over time, usually in a noninvasive recording via skin electrodes.

An electrocardiogram is obtained by measuring electrical potential between various points of the body using a biomedical instrumentation amplifier. A lead records the electrical signals of the heart from a particular combination of recording electrodes which are placed at specific points on the patient's body.

There are two types of leads – unipolar and bipolar. The former has an indifferent electrode at the center of the Einthoven's triangle (which can be likened to the "neutral" of a wall socket) at zero potential. The direction of these leads is from the "center" of the heart radially outward. These include the precordial (chest) leads and augmented limb leads – VR, VL, and VF. The bipolar type, in contrast, has both electrodes at some potential, with the direction of the corresponding lead being from the electrode at lower potential to the one at higher potential, e.g., in limb lead I, the direction is from left to right. These include the limb leads – I, II, and III.

2.10.1 Leads

A lead refers to a combination of electrodes that forms an imaginary line in the body along which the electrical signals are measured. The standard 12-lead ECG consists of 10 electrodes.

2.10.2 Limb Leads

Leads I, II, and III are the so-called limb leads because at one time, the subjects of electrocardiography had to literally place their arms and legs in buckets of salt water in order to obtain signals for Einthoven's string galvanometer. They form the basis of what is known as Einthoven's triangle.

Lead I is a dipole with the negative electrode on the right arm and the positive electrode on the left arm. Lead II is a dipole with the negative electrode on the right arm and the positive electrode on the left leg. Lead III is a dipole with the negative electrode on the left arm and the positive electrode on the left leg.

2.10.3 Augmented Limb

Augmented limb leads (aVR, aVL, and aVF) are derived from the same three electrodes as leads I, II, and III. However, they view the heart from different angles (or vectors) because the negative electrode for these leads is a modification of Wilson's central terminal, which is derived by adding leads I, II, and III together and plugging them into the negative terminal of the EKG machine. This zeroes out the negative electrode and allows the positive electrode to become the "exploring electrode" or a unipolar lead.

2.10.4 Precordial Leads

The precordial leads (V1, V2, V3, V4, V5, and V6) are placed directly on the chest. Because of their close proximity to the heart, they do not require augmentation. Wilson's central terminal is used for the negative electrode, and these leads are considered to be unipolar. The precordial leads view the heart's electrical activity in the so-called horizontal plane. The heart's electrical axis in the horizontal plane is referred to as the Z axis.

Leads V1, V2, and V3 are referred to as the right precordial leads and V4, V5, and V6 are referred to as the left precordial leads.

Lead V1 is placed in the fourth intercostal space to the right of the sternum. Lead V2 is placed in the fourth intercostal space to the left of the sternum. Lead V3 is placed directly between leads V2 and V4. Lead V4 is placed in the fifth intercostal space in the midclavicular line (even if the apex beat is displaced). Lead V5 is placed horizontally with V4 in the anterior axillary line. Lead V6 is placed horizontally with V4 and V5 in the midaxillary line.

2.10.5 Ground

An additional electrode (usually green) is present in modern four-lead and twelve-lead ECGs. This is the ground lead and is placed on the right leg by convention, although in theory it can be placed anywhere on the body. With a three-lead ECG, when one dipole is viewed, the remaining lead becomes the ground lead by default.

References

1. Wilson JS (2005) Sensor technology handbook. Handbook, Elsevier Inc, Burlington, MA
2. Ripka P, Tipek A (2007) Modern sensors handbook. ISTE Ltd, UK
3. Sinclair IR (2001) Sensors and transducers, 3rd edn. Elsevier
4. Buck RP, Lindner E, Kutner W, Inzelt G (2004) Piezoelectric chemical sensors. Pure Appl Chem 6:1139–1160

Chapter 3
Software Issues in Wireless Sensor Networks

Aleksandar Crnjin

As is the case with all other computing devices, sensor nodes too have to be programmed in order to be able to do something useful. Programming sensor nodes falls somewhere between programming desktop computers and small embedded devices: There is only conceptual similarity between programming a home computer with 2 GHz CPU and gigabytes of RAM and storage, and programming a tiny device with just a dozen kilobytes of RAM and an 8-bit processor running on just a few kilohertz. Still, programming sensor networks is not all that similar to embedded devices either: Programming embedded systems typically amounts to writing an assembly program and loading it into programmable ROM of the embedded device, while a sensor node usually has an operating system (such as TinyOS or ConTiki) which provides support for basic I/O operations, operation of the radio transceiver, and so on (Fig. 3.1).

Further difference is that sensor nodes are not conceived as being able to do anything useful autonomously; they always operate as a part of a sensor network. Having this in mind, programming paradigms for sensor networks have appeared which allow a programmer to program the sensor network as a whole.

We will now introduce a taxonomy using which we can identify three different approaches to programming sensor networks.

- **Full Custom** approach, in which we program a sensor node as if it were an embedded device with no OS
- **Scaled Traditional** approach, in which a sensor node is programmed the way an ordinary computer system is: using an operating system to operate the peripherals, putting the hardware aspects below the abstraction boundary, and thus enabling the programmer to put more effort into algorithms
- **Spatial Programming** approach, where we use the *spatial programming* paradigm so as to program the whole network as a unique entity, putting the internode communication aspects below the abstraction boundary as well

A. Crnjin (✉)
School of Electrical Engineering, University of Belgrade, Serbia
e-mail: acrnjin@gmail.com

L. Gavrilovska et al. (eds.), *Application and Multidisciplinary Aspects of Wireless Sensor Networks*, Computer Communications and Networks,
DOI 10.1007/978-1-84996-510-1_3, © Springer-Verlag London Limited 2011

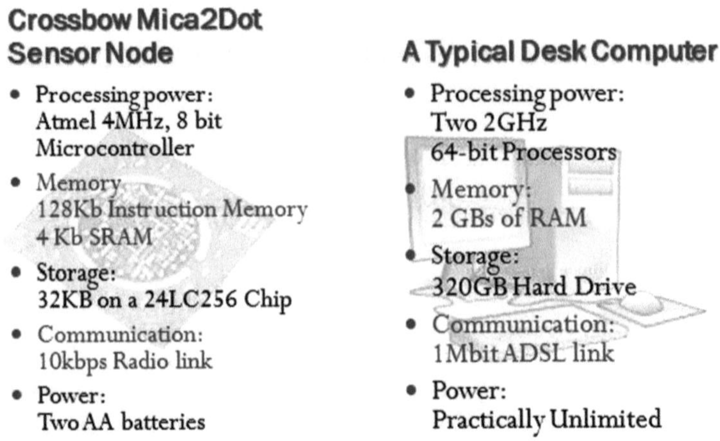

Fig. 3.1 A comparison of sensor nodes and personal computers

3.1 The Full Custom Approach

The "full custom approach" – programming a sensor node as if it was an ordinary embedded device – will not be described here in detail; we will rather concentrate on the pros and cons of this approach when used for sensor networks.

The *full custom software approach* means that we do not use an operating system; we program the sensor node's processor directly, using assembly language. Another option is to use a microcontroller-adapted high-level language (such as Keil C). Either way, we would have to operate the peripherals directly, using control and status registers, rather than primitives provided by an OS. The completed program would then be loaded into a sensor node's ROM, and executed when the device is powered on. With the *full custom hardware approach,* we go one step further and design the custom hardware for the application, rather than use readymade sensor modes.

A program developed with "full custom" would probably be the fastest and would consume the least battery power, as the whole functionality is optimized for the particular application which is developed. However, programming itself would be quite hard, and even attaining basic functionality of the application would take a long time.

3.2 The Scaled Traditional Approach

With the "scaled traditional approach" the sensor node is programmed with the same philosophy as when programming the personal computer: OS primitives are used to achieve hardware abstraction. Still, due to scarce, *scaled* resources of a sensor node, some compromises are required, most notably the need to conserve energy resources by providing an implicitly energy-efficient programming model.

The first obstacle we have to overcome when using this approach is to choose the right operating system for the sensor node. All conventional OSes for embedded systems (such as Windows CE or PalmOS) require a ROM capacity of a 100 kb or more; sensor nodes, however, typically have only a few kilobytes. The most popular solution today is the TinyOS, developed by U.C. Berkeley. Another option is the Java Squawk virtual machine, developed by Sun Microsystems for their Sun SPOT project.

3.2.1 The TinyOS Operating System

In order to facilitate the "scaled traditional" approach, a specialized operating system had to be devised; one that would provide the necessary primitives to operate the sensor node hardware, but would also cope with the quite limited resources of a sensor node. These limitations include

- **Not enough memory for stack**. All TinyOS programs have to operate within a single context, as it is impossible to perform traditional context switching, due to a very small amount of memory available for stack. This also means that TinyOS programs cannot rely on registers to save state.
- **Limited amount of memory**. This means that what available memory there is has to be allocated carefully. Dynamic allocation of memory is prohibited; the TinyOS *components* are arranged into a *configuration* at compile-time, and individual components each get their preassigned portion of memory, the *memory frame.*
- **Limited amount of energy**. Special care is taken to ensure that the battery power is conserved as much as possible. Busy waiting and interrupt polling is prohibited in all TinyOS compatible devices. TinyOS programs execute only in response to *events*; this is called the *event-driven* programming model.

3.2.1.1 TinyOS Component Model

To address these issues, the *TinyOS Component Model* was devised [1]. The parts of code in this model are organized into distinct entities based on functionality. For example, there may be an entity for operating the integrated radio-transceiver unit. (In fact, there is such an entity in TinyOS, the **rfm** system component.) These entities are associated with a statically allocated memory portion (called *memory frame)* and interconnected with other such entities through *interfaces.* Such completely described entities are called *components* (Fig. 3.2).

The *memory frame* is allocated at compile-time, based on total memory requirements of the component.

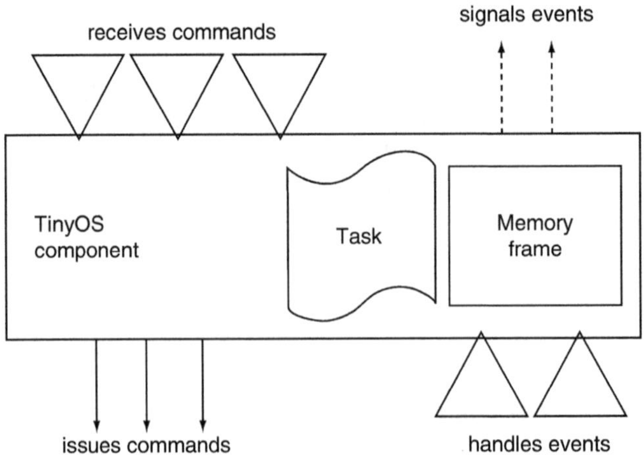

Fig. 3.2 Block view of a TinyOS component

The *interface* consists of registered commands and events, through which components communicate. The main difference between the two is in the fact that events propagate upward in the connection chain (from device driver components, such as aforementioned **rfm**, through bottom-level user components up to the top-level user components) and commands propagate downward in the connection chain. One component may choose to implement several interfaces, each consisting of a number of commands and events. More on this will be mentioned later, in the chapter on component interconnection.

The *code* contained in an individual TinyOS component consists of

- **Command handling routines**, which they execute in response to *commands* issued from other connected components
- **Event handling routines**, which they execute in response to *events* signaled by connected components
- **Tasks**, which components themselves schedule for later execution

3.2.1.2 System Components and Predefined Events

TinyOS comes with a certain number of *system components* which perform the function of device drivers. Examples include the already mentioned **rfm** transceiver-operating component [2], **Photo** and **Temperature** – the components for manipulating photo and temperature sensors, and so on.

In addition to predefined system components, there is also a predefined interface **StdControl,** consisting of events **Init, Start,** and **Stop**. Each TinyOS component has to implement this interface by providing handlers. For example, an **Init** event handler defines the response of the component to the event of system initialization. A component could use this handler in order to initialize its state to initial values. Actual execution of a program begins with handling of the **Start** event, in the mandatory **Main** component.

3.2.1.3 Component Configurations

Components are either **modules** or **configurations**. Each TinyOS "program" is a configuration of interconnected modules; each module is an encapsulation of code based on functionality, reminiscent of classes in object-oriented programming languages. Each configuration and module is described in a separate **.nc** file. Modules are connected through *interfaces*. An interface is a collection of events a given component can signal or be notified of, and commands that a component can issue or obey. Interfaces are also defined in their separate **.nc** files. A configuration is connected by "software wiring" of the interfaces. This is accomplished in two steps:

1. Each module lists the desired interface in either its **uses {...}** block, or its **provides {...} block**
2. A special line in the configuration code is added, which establishes a unidirectional link between an *interface provider* and an *interface user*. This is achieved with the following line of code:

```
User.UserInterfaceName -> Provider.ProvInterfaceName
```

or alternetively:

```
Provider.ProvInterfaceName <- User.UserInterfaceName
```

(The UserInterfaceName part may be omitted, if the interface names are the same.)

One example of an interface is the Timer interface, through which the system component for the onboard timer notifies the user component of regularly spaced *tick events*. The description of this interface is given in Timer.nc:

```
interface Timer {
  command result_t start(char type, unit32_t interval);
  command result_t stop();
  event result_t fired();
}
```

From the source code, it should be clear that the interface specifies one event – fired() – through which the Timer provider notifies the user of timer ticks, and accepts two commands – start, which accepts the type and interval parameters and starts the timer, and stop(), which stops the timer.

The actual semantics of the words *provider* and *user* might not be immediately clear: The events and commands which are the part of the interface are specified in the interface's **.nc** file. If a module declares itself to be a provider of a given interface, then it is responsible to implement all commands listed by the interface, and acquires the right (which it may or may not use) to signal events to those users of said interface to which it is connected through wiring. Likewise, if a module declares itself to be a user of an interface, then it is responsible to implement event handlers for every single event specified by the interface, and it attains the right to issue commands to the provider of the used interface.

In other words, the provider can signal events to users, and users can issue commands to the provider.

3.2.1.4 NesC "Hello World" Program: Blink

To give an overview of the issues involved, we consider a simple application that comes bundled with the TinyOS distribution, called **Blink**. As its name suggests, the Blink application uses the on-board timer to periodically change the state of a sensor node's LED, producing a blinking effect. The Blink application consists of four connected components: Main, BlinkM, SingleTimer, and Leds, as seen in the configuration declaration (Blink.nc):

```
configuration Blink {
}
implementation {
  components Main,BlinkM, SingleTimer, LedsC;
  // connecting Main toSingleTimer and BlinkM through StdControl
  Main.StdControl -> SingleTimer.StdControl;
  Main.StdControl -> BlinkM.StdControl;
  BlinkM.Timer -> SingleTimer.Timer;   //connecting BlinkM to
SingleTimer
  BlinkM.Leds -> LedsC;               //connecting BlinkM to LedC
}
```

In this example, we can note the following:

- Components Main, BlinkM, SingleTimer, and LedC are used. Code for these components is elaborated in separate files (for example, BlinkM code is found in BlinkM.nc).
- In the last four lines before the ending bracket, we can see how these components are interconnected – through "wiring" of their interfaces, using the notationUser.Interface -> Provider.Interface.
- BlinkM, the central module of the application, provides the StdControl interface, of which the implicit component Main is the user. The execution of the program "begins" with Main issuing the commands "Init" and "Start" to BlinkM. In response to these commands, BlinkM will start the timer, through the Timer interface. As BlinkM is notified of timer tick events, it toggles the LED, by issuing commands to the LedC component through the Leds interface.

```
//Implementation for Blink application:
// the red LED is toggled whenever Timer fires.
module BlinkM {
  provides {
     interface StdControl;
 }
 uses {
   interface Timer;
   interface Leds;
 }
}
implementation {
  //Handling of the Init command, issued by Main
  // just pass the Init command on to the Leds component
 command result_t StdControl.init() {
```

```
  call Leds.init();
  return SUCCESS; // SUCCESS is always returned
 }
 // Handling of the Start command, issued by Main:
 // set the rate for the clock component.
 command result_t StdControl.start() {
 // Start a repeating timer that fires every 1000ms
    return call Timer.start(TIMER_REPEAT, 1000);
 }
 // Handling of the Stop command, issued by Main:
 // stop the timer
 command result_t StdControl.stop() {
    return call Timer.stop();
 }
 // Handling the Fired event, issued by Timer:
 // issue redToggle command to Leds
 event result_t Timer.fired(){
    call Leds.redToggle();
    return SUCCESS;
 }
}
```

The configuration of the Blink application can be summarized in the following diagram (Fig. 3.3).

3.2.1.5 TinyOS Radio Communication

To facilitate efficient radio communication, TinyOS introduces the concept of "Active Messages." Active Messages effectively multiplex the access to the radio,

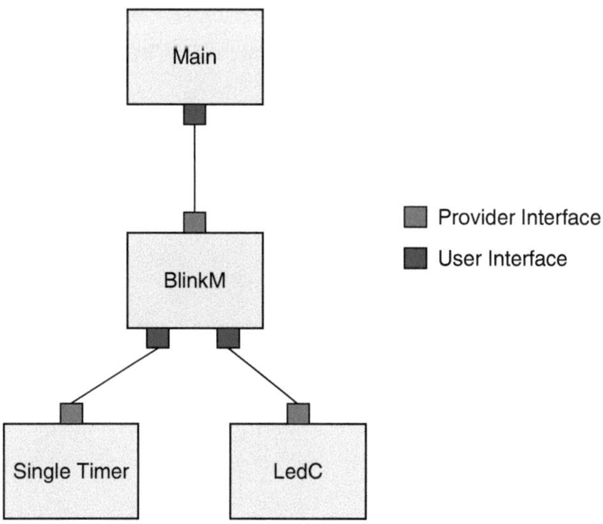

Fig. 3.3 Configuration of the Blink application

which is important in situations where multiple components demand access to the single physical radio medium at the same time. To this end, AM addresses are introduced: an AM message carries a "destination AM" field, in order to identify the node which should receive the message.

TinyOS uses two special interfaces to enable components to use Active Messages: **AMSend** and **AMPacket**. The physical format of the messages is defined in the **message_t** abstract type.

TinyOS uses a split-phase operation to send messages: a send message command (AMSend.send(…)) merely initializes the message sending. This command call returns SUCCESS, if the attempt to gain control over the radio has been successful and radio has begun sending. The programmer is now responsible not to issue any new requests to transmit on the radio (because calls for such requests will not have resulted in a SUCCESS and sending of a new message would not have started) until an event *AMSend.sendDone* is signaled.

Therefore, a general template for a message sending code looks like this:

```
typedef nx_struct MyCustomMsg {
  // insert message fields here
} MyCustomMsg;


implementation {
  bool busy = FALSE ;
  message_t packet ;
  … int DEST_ADDR = ... // specify destination for the message
  …
  /* inside some event handler, command handler or a task */
   if (! busy ()) {
     MyCustomMsg* mycustmsg = (MyCustomMsg*)
        (call Packet.getPayload(&packet, NULL));
    // fill mycustmsg's fields here
    if (call AMSend.send(DEST_ADDR, &packet, sizeof(MyCustomMsg))
== SUCCESS)
     {
         busy = TRUE; // prevent future req for radio
communication
       }
     }
…

// event handler for a sendDone event

event void AMSend.sendDone(message_t* msg, error_t error) {
  if (&pkt == msg) {
    busy = FALSE; // allow future requests for radio
communication
     }
   }
}
```

Receiving messages is realized through writing a handler for Receive.receive event:

```
event message_t* Receive.receive
```

```
  (message_t* msg, void*payload, uint8_t len) {
  ...
}
```

When a message is received from the transceiver, TinyOS calls this event handler and provides it with a pointer to the whole message, a pointer to its actual payload, and information about the message's length.

3.2.1.6 TinyOS: Conclusion

Programming in TinyOS is much simpler and quicker compared to the Full Custom model of assembly programming, and arguably easier, too. Still, the TinyOS programming model has suffered criticism for its still steep learning curve, especially for programmers accustomed to programming in established languages (such as Java).

3.2.2 Sun SPOTs and the Squawk Virtual Machine

The Squawk virtual machine, created by Sun Microsystems for their Sun SPOT technology, is one attempt at providing easier-to-use programming model to developers for sensor networks. Using Sun SPOTs and Squawk, a sensor network developer can write applications in a slightly modified versionof Java [3].

Squawk itself is a Java virtual machine running directly on SPOT hardware, without an underlying operating system. To facilitate execution of such sophisticated software, Sun SPOTs had to be designed with significantly more powerful hardware, compared to the TinyOS nodes (a Sun SPOT has a 180 MHz 32-bit processor). This might mean greater energy consumption and less battery life, compared to TinyOS nodes.

3.2.2.1 Programming Sun SPOTs

As mentioned already, Sun SPOTs are programmed using a variation of Java language. SunSPOT applications are MIDlets; the main class of the SunSPOT application extends the javax.microedition.midlet.MIDlet class, so constructs typical for MIDlets, such as startApp(), pauseApp(), and destroyApp(), form the skeleton of a Sun SPOT application. Therefore, an "entry point" for a Sun SPOT application is the startApp() method.

The whole functionality of the SPOT is abstracted using the EDemoBoard class. Typically, a programmer will obtain an instance of this class using EDemoBoard. getInstance() in the body of the startApp() method. Inputs (sensors and switches) and outputs (LEDs) are then reachable through Java interfaces, which are obtained using demoBoardInstance.getxxx () methods. For example, a temperature reading can be obtained using:

```
ITemperatureInput ourTempSensor = EDemoBoard.getADCTemperature();
double celsiusTemp = ourTempSensor.getCelsius();
double fahrTemp = ourTempSensor.getFahrenheit();
```

3.2.2.2 A SunSPOT "Blink" Application

As for TinyOS/nesC, we now present a Blink (Hello World) application for Sun SPOTs.

```
package org.sunspotworld;

importcom.sun.spot.sensorboard.EDemoBoard;
importcom.sun.spot.sensorboard.peripheral.ISwitch;
importcom.sun.spot.sensorboard.peripheral.ITriColorLED;
importcom.sun.spot.util.*;

importjavax.microedition.midlet.MIDlet;
importjavax.microedition.midlet.MIDletStateChangeException;

public class Blink extends MIDlet {
  private ITriColorLED [] leds =
eDemoBoard.getInstance().getLEDs();

  protected void startApp() throws MIDletStateChangeException {
    System.out.println("Hello, world");

    ISwitch sw1 =
EDemoBoard.getInstance().getSwitches()[EDemoBoard.SW1];
    leds[0].setRGB(100,0,0);        // set color to moderate red
    while (sw1.isOpen()) {          // done when switch is pressed
        leds[0].setOn();            // Blink LED Utils.sleep
        (250);                      // wait 1/4 seconds
        leds[0].setOff();
        Utils.sleep(1000);          // wait 1 second
 }
 notifyDestroyed();                 // cause the MIDlet to exit
}

 protected void pauseApp() {
     // This is not currently called by the Squawk VM
 }

 protected void destroyApp(boolean unconditional) throws
 MIDletStateChangeException {
        for (int i = 0; i < 8; i++) {
             leds[i].setOff();
        }
    }
 }
```

3.2.2.3 Radio Stream and Datagram Connections

Sun SPOT SDK provides two ways to perform communication between two Sun SPOT devices: the **radiostream** protocol and the **radiogram** protocol. The basic difference is in guarantees of delivery: the radiostream protocol provides stream-based reliable service, while the radiogram protocol makes no guarantees on delivery or ordering.

In the radio stream case, individual SPOT devices are all assigned URLs of the format

```
radiostream://<DestAddress>:<PortNumber>
```

Where DestAddress is a 64-bit IEEE address of the remote SPOT, and PortNumber is an integer number in the range [0,255] which serves as the identifier of the port for this particular connection. Once a connection is opened, through a call to the Connector.open() method, the programmer is able to communicate with the remote SPOT in the standard Java fashion, using DataStreams:

```
RadiostreamConnection conn =
   (RadiostreamConnection)
Connector.open("radio://00AB.CC21.00A0.0006:100");
DataOutputStream dos = conn.openDataOutputStream();

try{
   dos.writeUTF("A message");

   dos.flush();
} catch (NoRouteException e) {
  System.out.println ("00AB.CC21.00A0.0006 is unreachable");
} finally {
dos.close();
conn.close();
```

In order for this example to work, the remote SunSPOT at 00AB.CC21.00A0.0006 must also open a connection to the calling SunSPOT, on the same port (100). It should also open a DataInputStream, and issue a call to readUTF().

In the case of the radiogram protocol, the situation is a little bit different. There is a difference in how the connection is opened on the client side and on the server side (the server listens for requests on the specified port, while the client issues a request using the server's URL and the same specified port. The URLs are of the form

```
radiogram://<Address>:<Port>
```

The data itself has to be packed into datagrams, prior to sending:

```
RadiogramConnection conn =
(RadiogramConnection)Connector.open("radiogram://
00AB.CC21.00A0.0006:10");
Datagram dg = conn.newDatagram(conn.getMaximumLength());
try {
  dg.writeUTF("A message");
  conn.send(dg);
} catch (NoRouteException e) {
  System.out.println ("00AB.CC21.00A0.0006 is unreachable ");
} finally {
  conn.close();
}
```

3.2.2.4 Sun SPOTs/Squawk: A Conclusion

Sun SPOTs certainly provide an easier programming model and as a result they are very popular within certain educational circles. However, their bulky size (compared to TinyOS-based platforms), higher price (for noneducational use), and higher battery consumption still make them somewhat impractical for some industrial applications.

3.3 Spatial Programming – A Brief Overview

Recently, there has been a lot of research effort into programming paradigms for spatial computing. These paradigms are applicable for sensor networks, as they too consist of small computing devices distributed over a spatial area.

Typical challenges which a sensor network programmer faces in the traditional programming model include:

- Inherent unreliability of both wireless sensor links (due to communication interference and noise) and sensor nodes (due to depleting battery power)
- Unsuitability of IP data-based addressing: geographical approaches allow for easier replacing of spent nodes, as it is enough to move into physical space of the malfunctioned node to continue receiving messages
- Too much overhead in setting up and maintaining connections between communi cating sensor nodes
- Resulting unscalability of solutions developed using the traditional programming model

Spatial Programming paradigms attempt to address these issues using concepts such as:

- **Shared spatial variables**, allowing programmers to read and write such variables directly from their code, without the need to implement synchronization and communication as they will be performed by the spatial programming framework
- **Location-based naming** for individual computing devices in a spatial scenario, instead of generic IP addresses
- **Smart messages**, through which both data, application state and code snippets can be exchanged between computing devices
- **Centralized programming**, writing code which describes the behavior of spatial application all in one place. Actual code snippets for execution on remote computing devices will be sent through smart messages

Spatial computing is an emerging, highly active, interesting research area. For more information and one example implementation, an interested reader can follow [4, 5].

References

1. Lewis P (2006) TinyOS Programming
2. TinyOS Documentation Wiki
3. SunSPOT Developer's Guide (2009) SUN Microsystems
4. Bachrach J, Beal J (2006) Programming a sensor network as an amorphous medium, DCOSS
5. Bachrach J, et al. (2008) Spatial approaches to pervasive computing, tutorial at IEEE SASO

Chapter 4
Position-Based Routing Protocols for Ad Hoc Networks

Liana K. Qabajeh, Laiha Mat Kiah, and Mohammad M. Qabajeh

Wireless Ad Hoc networks are collections of nodes that can communicate without any fixed infrastructure. A crucial problem in Ad Hoc networks is finding an efficient and correct route between a source and a destination. The need for scalable and energy-efficient protocols, along with the recent availability of small, inexpensive, and low-power positioning instruments justify adopting position-based routing algorithms in mobile ad hoc networks.

This chapter presents an extensive overview of the existing Ad Hoc routing protocols that make forwarding decisions based on the geographical position of a packet's destination. We outline the main problems for this class of routing protocols and a qualitative comparison of the existing methods is done. We conclude our findings by investigating opportunities for future research.

4.1 Introduction

Ad Hoc wireless networks are self-organizing multi-hop wireless networks, where all the hosts (or nodes) take part in the process of forwarding packets. Ad Hoc networks can quickly and inexpensively be set up as needed since they do not require any fixed infrastructure, such as base stations or routers. Therefore, they are highly applicable in many fields, such as emergency deployments and community networking.

Mobile nodes in an Ad Hoc network have limited radio transmission range. Nodes that are unable to communicate directly with each other require that intermediate nodes forward packets for them. The function of a routing protocol in Ad Hoc network is to establish routes between different nodes. A fundamental and challengeable task in Ad Hoc wireless network is an efficient routing protocol since all the nodes in the network act as hosts as well as routers.

L.K. Qabajeh (✉), L.M. Kiah, and M.M. Qabajeh
Computer Science and Information Technology Faculty, University of Malaya, Malaysia
e-mail: liana_tamimi@ppu.edu

L. Gavrilovska et al. (eds.), *Application and Multidisciplinary Aspects of Wireless Sensor Networks*, Computer Communications and Networks,
DOI 10.1007/978-1-84996-510-1_4,

Ad Hoc network routing protocols are difficult to design in general. There are two main reasons for that; the highly dynamic nature of the Ad Hoc networks due to high mobility of the nodes, and the need to operate efficiently with limited resources such as network bandwidth and limited memory and battery power of the individual nodes in the network.

Moreover, routing protocols in Ad Hoc networks, unlike static networks, do not scale well due to frequently changing network topology, lack of predefined infrastructure like routers, peer-to-peer mode of communication, and limited radio communication range [1].

For these reasons, many routing protocols that are compatible with the characteristics of Ad Hoc networks have been proposed. In general, they can be divided into two main categories: topology-based and position-based. Topology-based routing protocols use information about links that exist in the network to perform packet forwarding. They are, in turn, divided into three categories: proactive, reactive, and hybrid (hierarchical) protocols.

Proactive routing protocols periodically broadcast control messages in an attempt to have each node always know a current route to all destinations, and remove local routing entries if they time out. We observed that proactive routing protocols are less suitable for Ad Hoc wireless networks because they constantly consume power throughout the network, regardless of the presence of network activity. Also they are not designed to track topology changes occurring at a high rate.

On the other hand, reactive routing protocols are deemed more appropriate for wireless environments because they initiate a route discovery process only when data packets need to be routed. Many Ad Hoc routing protocols that use reactive route determination have been developed such as Ad Hoc On-demand Distance Vector (AODV) [2] protocol. One advantage of reactive routing protocols is that no periodic routing packets are required. However, they may have poor performance in terms of control overhead in networks with high mobility and heavy traffic loads. Scalability is said to be another disadvantage because they rely on blind broadcasts to discover routes.

As seen, proactive routing uses excess bandwidth to maintain routing information, while reactive routing involves long route request delays. Reactive routing also inefficiently floods the entire network for route determination. Hybrid routing protocols aim to address these problems by combining the best properties of both approaches. The disadvantage of ZRP is that for large values of routing zone the protocol can behave like a pure proactive protocol, while for small values it behaves like a reactive protocol [3]. In general, topology-based are considered not to scale in networks with more than several hundred nodes [4].

In recent developments, position-based routing protocols exhibit better scalability, performance, and robustness against frequent topological changes [1, 4]. Position-based routing protocols use the geographical position of nodes to make routing decisions, which results in improving efficiency and performance. These protocols require that a node be able to obtain its own geographical position and the geographical position of the destination. Generally, this information is obtained via Global Positioning System (GPS) and location services [5]. The routing decision at each node is then based on the destination's position contained in the packet and the position of the forwarding node's neighbors. So the packets are delivered to the nodes in a given geographic region in a natural way. There are different kinds of position-based protocols that are categorized into three main groups: restricted directional flooding, greedy, and hierarchical routing protocols [6] (to be discussed in Section 4.2).

This survey, which is a continuation of our work in [7], gives an overview of a large percentage of existing position-based routing protocols for mobile Ad Hoc networks. We outlined the main problems that have to be solved for this class of routing protocols and presented the solutions that are currently available. The discussed protocols are also compared with respect to the used location service, the used forwarding strategy, tolerability to position inaccuracy, robustness, implementation complexity, scalability, packet and processing overhead, guaranteeing loop-freedom, probability of finding the shortest path, and the suitable network density to be implemented in.

The protocols that have been selected for analysis are MFR [8], DIR [9], GPSR [10], ARP [1], IPBBLR [4], DREAM [11], LAR [12], LARWB [13], MLAR [14], GRID [15], TERMINODES [16], LABAR [17], POSANT [18], PAGs [19], SPAAR [5], AODPR [20], and SGF [21]. It worth noting that many other position-based routing protocols exist for mobile Ad Hoc networks; however, we have selected what we regard as representative for the existing approaches.

The rest of the chapter is organized as follows. Section 4.2 presents the basic idea and principles of position-based addressing and routing. Section 4.3 gives an overview of the selected position-based routing protocols. Section 4.4 contains a qualitative comparison of the discussed protocols. Directions of future research are discussed in Section 4.5. Finally, we conclude the chapter in Section 4.6.

4.2 Basic Principles of Position-Based Routing

The main prerequisite for position-based routing is that a sender can obtain the current position of the destination. Typically, a location service is responsible for this task. Existing location services can be classified according to how many nodes host the service. This can be either some specific nodes or all nodes of the network. Furthermore, each location server may maintain the position of some specific or all nodes in the network. The four possible combinations can be abbreviated as some-for-some, some-for-all, all-for-some, and all-for-all [6].

There are three main packet-forwarding strategies used for position-based protocols: greedy forwarding, restricted directional flooding, and hierarchical approaches. While their main objective is to utilize available position information in the Ad Hoc routing, their means to achieve it are quite different. Most position-based protocols use greedy forwarding to route packets from a source to the destination. Greedy protocols do not establish and maintain paths from source to the destination, instead, a source node includes the approximate position of the recipient in the data packet and selects the next hop depending on the optimization criteria of the algorithm;

the closest neighbor to the destination for example [1, 6]. Similarly, each intermediate node selects a next hop node until the packet reaches the destination. In order for the nodes to be enable to do this, they periodically broadcast small packets (called beacons) to announce their position and enable other nodes maintain a one-hop neighbor table [4, 9, 12]. Such an approach is scalable and resilient to topology changes since it does not need routing discovery and maintenance; however, periodic beaconing creates lot of congestion in the network and consumes nodes' energy [1, 4, 21].

While the beaconing frequency can be adapted to the degree of mobility, a fundam ental problem of inaccurate (outdated) position information always presents: a neighbor selected as a next hop may no longer be in transmission range. This leads to a significant decrease in the packet delivery rate with increasing node mobility. To reduce the inaccuracy of position information, it is possible to increase the beaconing frequency. However, this also increases the load on the network by creating lot of congestion, increasing the probability of collision with data packets, and consuming nodes' energy [1, 4].

Unfortunately, greedy routing may not always find the optimum route, even it may fail to find a path between source and destination when one exists [6, 21]. An example of this problem is shown in Fig. 4.1. Nodes' transmission range is shown in the figure.

Note that there is a valid path from S to D. The problem here is that S is closer to the destination D than any of the nodes in its transmission range; therefore, greedy forwarding will reach a local maximum from which it cannot recover. Generally, greedy forwarding works well in dense networks, but in sparse networks it fails due to voids (regions without nodes) [1, 9].

In restricted directional flooding, the sender will broadcast the packet (whether the data packet or route request packet) to all single-hop neighbors toward the destination. The node which receives the packet, checks whether it is within the set of nodes that should forward the packet (according to the used criteria). If yes, it will retransmit the packet. Otherwise, the packet will be dropped. In restricted directional flooding, instead of selecting a single node as the next hop, several nodes participate in forwarding the packet in order to increase the probability of finding

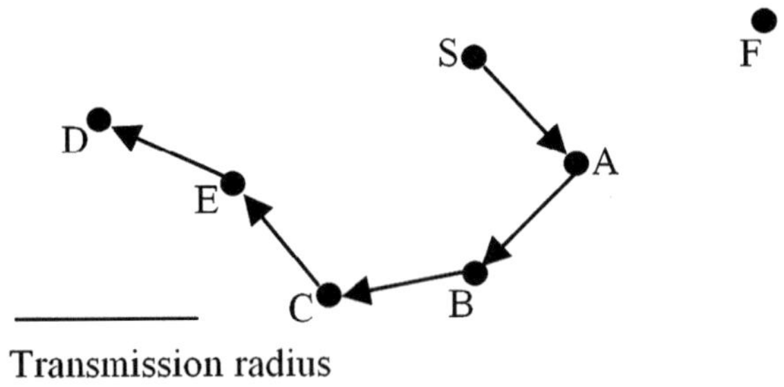

Fig. 4.1 Greedy routing failure example

the shortest path and are robust against the failure of individual nodes and position inaccuracy.

The third forwarding strategy is to form a hierarchy in order to scale to a large number of mobile nodes. Some strategies combine nodes location and hierarchical network structures by using the zone-based routing such as LABAR. Others use the dominating set routing such as GRID. Some others, such as TERMINODES, present a two-level hierarchy within them; if the destination is close to the sender (in number of hops), packets will be routed based on a proactive distance vector. Greedy routing is used in long-distance routing; therefore, they have characteristics similar to those of greedy forwarding.

4.3 Overview of Selected Position-Based Routing Protocols

In this section the selected protocols are described. For each protocol, we tried to summarize its main objectives, how it works, and its advantages and disadvantages compared to other protocols.

4.3.1 MFR

Some greedy position-based routing protocols, such as Most Forward within distance R (MFR) [8], aim to minimize the number of hops by selecting the node with the largest progress from the neighbors, where progress is defined as the projection of the distance of the next hop from the sender on the straight line between the sender and the destination [4, 7, 12]. In Fig. 4.2, if the MFR is used the source S will choose the node A as the next hop since it has the largest progress to the destination D.

As other greedy forwarding protocols, MFR has the shortcomings of either not guaranteeing to find a path to the destination or finding a path which is much longer than the shortest path. Moreover, nodes periodically should broadcast beacons to announce their positions and enable other nodes maintain a one-hop neighbor table.

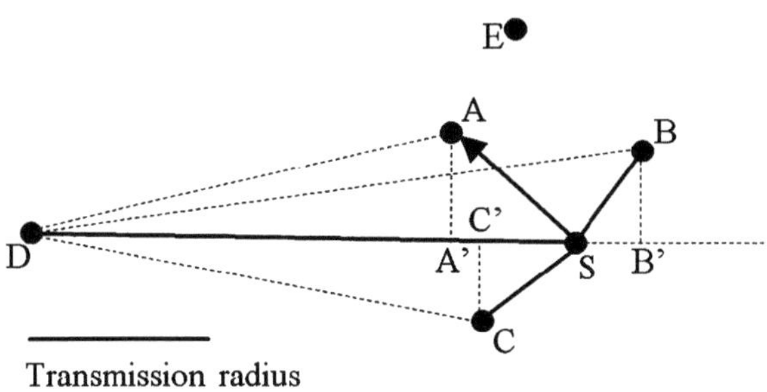

Fig. 4.2 MFR example

MFR is probed to be a loop-free algorithm [23] since it always forces message to make a step closer to the destination, and it is the only progress-based algorithm competitive in terms of hop count [24]. However, choosing the node with the largest progress as the next hop will increase the probability that the two nodes disconnected from each other before the packet reaches the next hop. So, the packet drop rate increases greatly, especially in highly mobile environments. Such a situation is very common due to neighbor table inconsistency [1].

4.3.2 DIR

Compass routing algorithms, such as DIR [9], try to minimize the spatial distance that a packet travels and base on forwarding the packet to the neighboring node that minimizes the angle between itself, the previous node, and the destination [6, 8, 12]. The source or intermediate node A uses the location information of the destination D to calculate its direction. Then the message m is forwarded to the neighbor C, such that the direction AC is closest to the direction AD. This process repeats until the destination is, eventually, reached [24]. Consider the network in Fig. 4.3, where the transmission radius is as indicated in the figure. The direction AC is closest to the direction AD among candidate directions AS, AB, AC, AG, and AF. So the path selected by DIR method is SACD.

As a greedy protocol, DIR suffers from congestion created by frequent beaconing and it may not always find the optimum route, even it may fail to find a path between source and destination if it exists. The DIR method, and any other method that includes forwarding message to neighbor with closest direction such as DREAM [11], is not loop-free as shown in [23] using the counterexample in Fig. 4.4. The loop consists of four nodes denoted S, B, C, and A. The transmission radius is as indicated in the figure. Let the source be any node in the loop, e.g., S. Node S selects node B to forward the message, because the direction of B is closer to destination D than the direction of its other neighbor A. Similarly node B selects C, node C selects A and node A selects S.

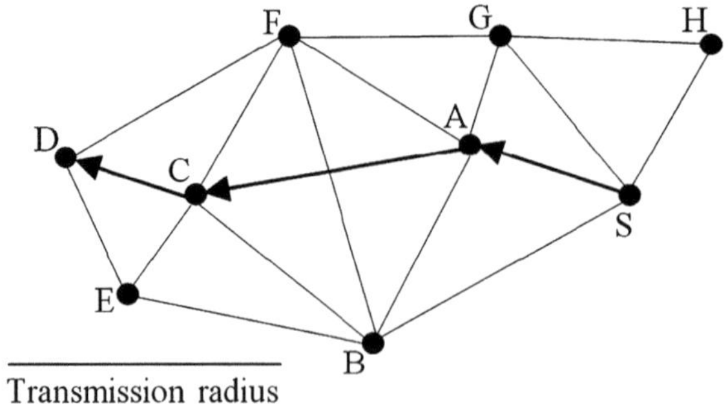

Fig. 4.3 DIR example

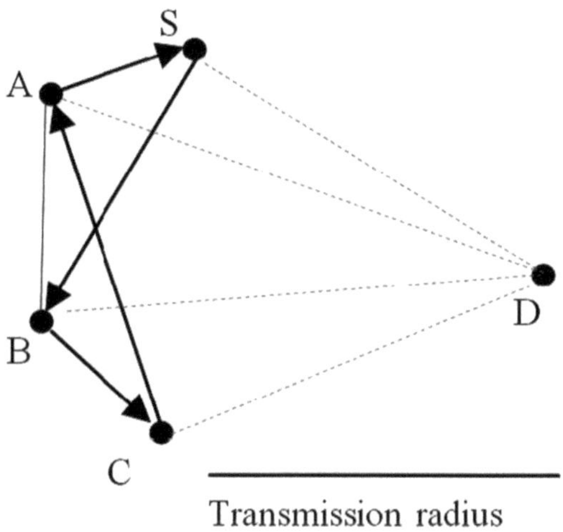

Fig. 4.4 A loop in directional routing

4.3.3 GPSR

Nearly Stateless Routing with Guaranteed Delivery are schemes where nodes maintain only some local information to perform routing. The face routing and Greedy-Face-Greedy (GFG) schemes were described in [25]. In order to ensure message delivery, the face routing (called perimeter algorithm in [10]) constructs planar and connected so-called Gabriel subgraph of the unit graph, and then applies routing along the faces of the subgraph (e.g., by using the right-hand rule) that intersect the line between the source and the destination. If a face is traversed using the right-hand rule then a loop will be created; since face will never be existed. Forwarding in right-hand rule is performed using directional approach. To improve the efficiency of the algorithm in terms of routing performance, face routing can be combined with algorithms that usually find shorter routes, such as the greedy algorithm to yield GFG algorithm [18, 24]. Routing is mainly greedy, but if a mobile host fails to find a neighbor closer than itself to the destination, it switches the message from "greedy" state to "face" state [6].

Authors in [10] transformed GFG algorithm into Greedy Perimeter Stateless Routing (GPSR) protocol by including IEEE 802.11 medium access control scheme. The perimeter routing strategy of the GPSR is based on planar graph traversal and proposed to address the local maximum problem of greedy forwarding [6, 23]. It is performed on a per-packet basis and does not require the nodes to store any additional information. A packet enters the recovery mode when it arrives at a local maximum. It returns to greedy mode when it reaches a node closer to the destination than the node where the packet entered the recovery mode [6]. GPSR guarantees that a path will be found from the source to the destination if there exists at least one such path in the original non-planar graph [6].

In [10] GPSR was experimented and compared with non-position-based protocol; Dynamic Source Routing (DSR) [26]. GPSR protocol consistently delivered over 94% data packets successfully; it is competitive with DSR in this respect on 50 node networks, and increasingly more successful than DSR as the number of nodes increases. The routing protocol traffic generated by GPSR was constant as mobility increased, while DSR must query longer routes with longer diameter and do so more often as mobility increases. Thus, DSR generates drastically more routing protocol traffic in simulations with over 100 nodes [10]. Therefore, the scalability seems to be the major advantage of this class of algorithms over source-based protocols. However, these simulations did not include the traffic and time required to look up the position of the destination. It was also assumed that the position of the destination is accurately known by the sender [6].

Nearly stateless schemes are likely to fail if there is some instability in the transmission ranges of the mobile host, when the network graph includes nodes with irregular transmission ranges [17, 23]. Instability in the transmission range means that the area a mobile host can reach is not necessarily a disk. This unstable situation occurs if there are obstacles (e.g., buildings, bad weather) that disrupt the radio transmission [24]. In GPSR, as other greedy forwarding protocols, periodic beaconing creates lot of congestion in the network and consumes nodes' energy. In addition, GPSR uses link-layer feedback from Media Access Control (MAC) layer to route packets; such feedbacks are not available in most of the MAC layer protocols [1]. Finally, planarizing the underlying graph (network) is computationally expensive and requires up-to-date neighborhood information [1].

4.3.4 ARP

Another scalable position-based routing protocol is Angular Routing protocol (ARP) [1]. In ARP, nodes emit a hello packet on need-basis (non-periodic) at a rate proportional to their speeds. These hello packets enable that each node maintains a one-hop neighbor table. ARP uses geographic forwarding to route packets to the destination. If the geographic forwarding fails, it uses an angle-based forwarding scheme to circumvent voids in sparse networks. ARP does not need any link-layer feedbacks like GPSR.

If a source wants to send a packet to a specific destination, it selects the geographically closest node towards to the destination among its neighbors as the next hop. Each intermediate node follows this next hop selection criterion. Thus, at each hop the packet progresses toward the destination by a distance ≤0.9R, where R is the radio range of the node. This is done to avoid the problem of leaving the next hop node out from the transmission range of the current node.

If no node is closer to the destination than the source node, or any intermediate node, then the node selects a neighboring node that makes minimum angle, among available neighbors. Figure 4.5 shows the angle-based forwarding to circumvent

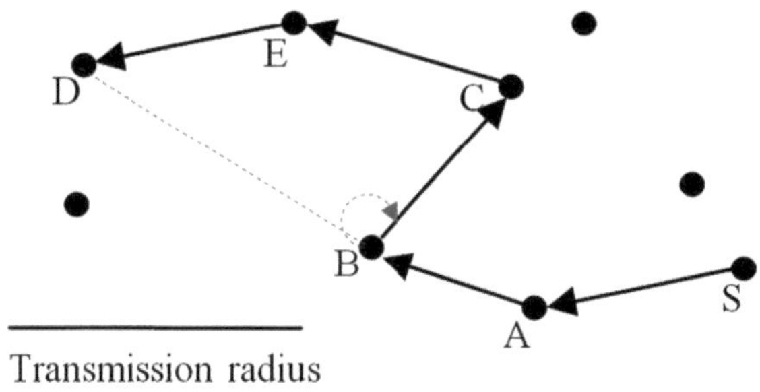

Fig. 4.5 Angle-based forwarding to circumvent voids in ARP

voids. The intermediate node B has no neighbors closer to the destination D than itself. In such a situation B selects a next hop that is making minimum angle toward destination, i.e., node C. After selecting a next hop node, the intermediate node appends its ID to the packet header. For each data packet, ARP header memorizes a maximum of k last visited hops in order to avoid selecting a next-hop whose ID presents in the ARP header. This memorization technique helps ARP in avoiding local loops, but does not guarantee its loop freedom. It is clear that assigning k to a small value will decrease the percentage of local loops avoided. On the other hand, assigning it to a large value will enlarge the size of the packet which in turn will increase the packet overhead.

The simulations in [1] showed that ARP is scalable, achieves high packet delivery rate while incurring low overhead compared to GPSR. Emitting hello packets on need-basis reduces the problems associated with beaconing. Also, using the angle-based forwarding to circumvent voids increases the probability of finding a path (not necessarily the optimal one) in sparse networks.

4.3.4.1 I-PBBLR

Most position-based routing protocols use forwarding strategies based on distance, progress, or direction. Improved progress Position Based BeaconLess Routing algorithm (I-PBBLR) [4] combines the traditional progress with the direction metric to form the improved progress definition. There are many methods to combine the progress with direction, such as weighted addition, simple multiplication. The authors have chosen the cosine of the angle since its value is between 0 and 1, and it is even. If the traditional progress is multiplied by the cosine of the angle, both the minimum and maximum of the progress are not changed. Also, it fits for the need that the node has smaller angle will forward packet earlier. Finally, they guarantee loop freedom as the packets are always forwarded a step closer to the destination.

I-PBBLR tries to eliminate the beaconing drawbacks by using a beaconless protocol. In beaconless protocols the sender makes nondeterministic routing decisions, implicitly allowing opportune receiving nodes to determine a packet's next hop through contention at transmission time. In I-PBBLR, if a source node has a data packet to send, it first determines the position of the destination, stores these geographical coordinates along with its own current position in the header of the packet, and broadcast the packet to all neighboring nodes (since it does not possess knowledge of neighboring nodes positions).

Nodes located within the forwarding area of the relaying node, apply Dynamic Forwarding Delay (DFD) prior to relaying the packet, whereas nodes outside this area drop the received packet. The value of the DFD depends on the relative position coordinates of current, previous, and destination node. Eventually, the node that computes the shortest DFD forwards the packet first by broadcasting it to all neighboring nodes after replacing the previous node's position by its current position in the header). Every node in the forwarding area detects the further relaying of the packet and cancels its scheduled transmission of the same packet. This mechanism allows selecting one neighbor as next hop in a completely distributed manner without having knowledge of the neighboring nodes, which is achieved by applying the concept of DFD. The simulation results showed that position-based beaconless routing using the improved progress reduced the overhead and increased delivery rate by 3–5% compared with using the traditional progress.

4.3.5 DREAM

Distance Routing Effect Algorithm for Mobility (DREAM) [11] is an example of restricted directional flooding routing protocols, that within them, the sender will broadcast the packet toward nodes in a limited sector of the network; to all single hop neighbors toward the destination [11, 12]. DREAM algorithm is a proactive protocol that uses a limited flooding of location update messages [24]. In DREAM, each node maintains a position database that stores position information about all other nodes in the network. Its location service can therefore be classified as an all-for-all approach. Thus, each node regularly floods packets to update the position information maintained by the other nodes. The higher the speed of a node the more the frequency with which it sends position updates. Also, the distance that a position update may travel before it is discarded provides accurate position information in the direct neighborhood of a node and less accurate information at nodes farther away, but this does not cause a problem since intermediate hops are able to update the position information contained in the data packet [6, 10]. In DREAM the message is forwarded to all neighbors whose direction belongs to the region that is likely to contain the destination D, called the expected region. Expected region is determined by the tangents from the source S to the circle centered at D and with radius equal to a maximal possible movement of D since the last location update [6, 23]. The neighboring hops repeat this procedure using their information on D's position.

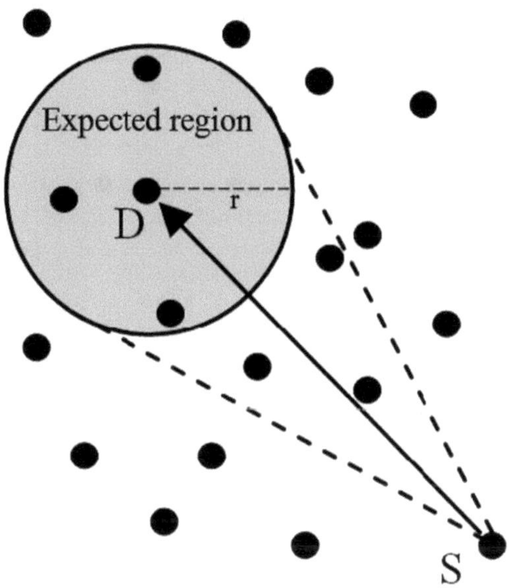

Fig. 4.6 Example of the expected region in DREAM

Figure 4.6 gives an example for the expected region in DREAM. If a node does not have a one-hop neighbor in the required direction, a recovery procedure has to be started. However, this procedure is not part of the DREAM specification [6].

Since DREAM uses the restricted directional flooding to forward data packets themselves, there will be multiple copies of each packet at the same time. This increases the probability of using the optimal path; however, it decreases its scalability to large networks with a high volume of data transmissions and makes it more suitable for applications that require a high reliability and fast message delivery for infrequent data transmissions.

4.3.6 LAR

Like DREAM, Location-Aided Routing (LAR) [12] is an example of restricted directional flooding routing protocols; however, partial flooding is used in LAR for path discovery purpose and in DREAM for packet forwarding [24]. Hence, LAR does not define a location-based routing protocol but instead proposes the use of position information to enhance the route discovery phase of reactive Ad Hoc routing approaches [6]. If no information is available in the source about the position of the destination, LAR is reduced to simple flooding [6]. Otherwise, the expected zone (the area containing the circle and two tangents) is fixed from the source and defined

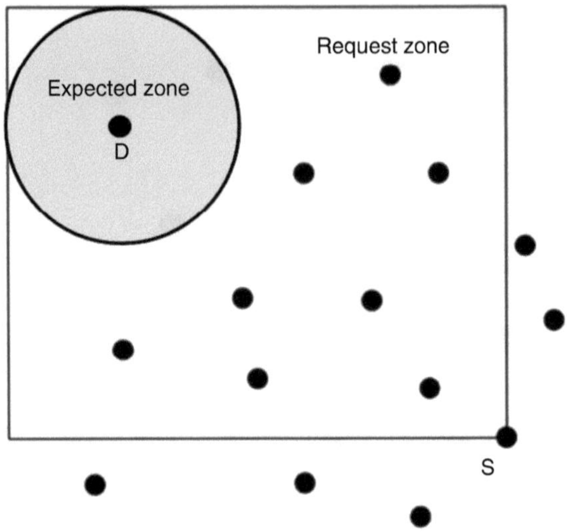

Fig. 4.7 Example of request and expected zones in scheme 1 of LAR

based on the available position information (e.g., from a route that was established earlier) [6, 23]. A request zone is defined as the set of nodes that should forward the route discovery packet. The request zone typically includes the expected zone. Two request zone schemes have been proposed in [12]. The first scheme is a rectangular geographic region. In this case, nodes will forward the route discovery packet only if they are within that specific region. This type of request zone is shown in Fig. 4.7.

In LAR scheme 2, the source or an intermediate node will forward the message to all nodes that are closer to the destination than itself. Thus, the node that receives the route request message will check if it is closer to the destination than the previous hop and it will retransmit the route request message; otherwise, it will drop the message [11, 12].

In order to find the shortest path in the network level, instead of selecting a single node as the next hop, several nodes are selected for managing the route request message and each of them will put its IP address in the header of the request packet. Therefore, the route through which the route request message is passed will be saved in the header of the message [13]; message size grows as it goes far from the source and the routing overhead will be increased. In LAR, if the discovered route breaks for any reason, route discovery process is started again.

4.3.7 LARWB

Routes in LAR are often broken due to mobility [14]. New routes must be rediscovered to continue routing of packets in the queue. This problem was solved by Location-Aided Routing With Backup (LARWB) [13] since another route is selected as a backup route which is used when a breakage appears in the primary route.

Selecting an appropriate backup route can be done by considering two points; the primary and the backup routes must have the minimum common nodes, the backup route should have a low probability of having nodes that may leave the radio range of their previous hop node. Experimental results in [13] showed that by using LARWB, the number of nodes which participate in routing operation, the average number of exchanged messages in route discovery process and also the average time of route discovery are reduced considerably.

4.3.8 MLAR

Multipath Location Aided Routing (MLAR) [14] is a multipath routing version of LAR that works efficiently in both 2-Dimensional (2D) and 3-Dimensional (3D) networks. By multipath they meant caching of alternate paths to be used on failure of the primary path and not the use of multiple simultaneous paths which can lead to out-of-order packet delivery problems. They cached the two most recently received routes even if they are longer. They believe that the most recently received path (even if it is longer) is the path most likely to succeed since mobility is more likely to break an older path. However, a routing protocol with longer average hop counts may have lower packet delivery rate. This is because the probability of a packet being dropped is higher when packets traverse longer paths than shorter paths. If the second path also failed a new route request cycle is initiated. Since the packet header contains the entire source route, all paths are checked easily as being loop-free at each node that stores routes.

In order to be able to compare MLAR to other existing protocols, the authors have extended ns-2 to support 3D mobility models and routing protocols. The simulation results demonstrated the performance benefits of their multipath position-based algorithm over a multipath non-position-based algorithm, Ad Hoc On-demand Multipath Distance Vector routing (AOMDV) [27], as well as with both their single path versions (LAR and AODV) in both 2D and 3D. Only AOMDV consistently performs better than MLAR in terms of overall packet delivery, but does so at the cost of more frequent flooding of control packets and thus uses significantly more bandwidth. Hence, MLAR has lower bandwidth and energy usage than non-position-based algorithms and is more scalable and efficient. Moreover, MLAR performs consistently better than LAR in terms of packet delivery ratio by as much as 30% in some cases.

4.3.9 GRID

The two main strategies used to combine nodes location and hierarchical network structures are the zone-based routing and the dominating set routing [24]. In GRID algorithm [15] the dominating set concept is applied. A set is dominating if all the nodes in the system are either in the set or neighbors of nodes in the set. Routing based on a connected dominating set is a promising approach, since the searching space for a route is reduced to nodes in the set. GRID tries to exploit location information in route discovery, packet relay, and route maintenance.

In GRID the geographic area is partitioned into a number of squares called grids. In each grid, one mobile host (the one nearest to the physical center of the grid) will be elected as the leader of the grid. The size of each grid depends on transmission radius R, and several options are proposed, with general idea of one leader being able to communicate directly with leaders in neighboring grids, and all nodes within each grid being connected to their leaders. Routing is then performed in a grid-by-grid manner through grid leaders, and non-leaders have no such responsibility. Hence, the number of packets related to route search is insensitive to the network density. On the contrary, the cost slightly goes down as the host density increases, since routes are becoming more stable with denser hosts.

In GRID, efforts are made in two directions to reduce the route search cost; using the locations of source and destination to confine the search range (like request zone in LAR) and delegating the searching responsibility to the gateway hosts. One attractive feature of GRID is its strong route maintenance capability since when a leader moves, another leader from the same grid replaces it by a handoff procedure. The probability of route breakage due to a nodes roaming is reduced since the next hop is identified by its physical location, instead of its address. Grid uses a specific field to detect duplicate request packets from the same source, so endless flooding of the same request can be avoided, i.e., it is loop-free routing.

Simulations in [15] showed that GRID can reduce the probability of route breakage, reduce the number of route discovery packets and lengthen routes' lifetime. On the other hand their simulations showed that GRID uses longer paths than that used with LAR, since the former always confines relay hosts to gateway hosts while LAR tries to search the route with the smallest host count. Also, the authors do not elaborate on route maintenance required when a grid remains empty after its leader and only node leaves it [24]. Feeney and Nillson in [28] and Shih et al. in [29] concluded that the idle power consumption is nearly as large as that of receiving data. Also, a node in idle mode spends about 15–30 times more energy than if it is in sleep mode. Therefore, developing protocols that have as many as possible sleeping nodes, such as GRID, will save network energy significantly.

4.3.10 TERMINODES

TERMINODES [16] is an example of hierarchical routing protocols. TERMINODES presents a two-level hierarchy within which, if the destination is close to the sender (in terms of number of hops), packets will be routed base on a proactive distance vector. Greedy routing is used in long distance routing [13]. TERMINODES addresses the following objectives: scalability (both in terms of the number of nodes and geographical coverage), robustness, collaboration, and simplicity of the nodes [24].

This routing scheme is a combination of two protocols called Terminode Local Routing (TLR) and Terminode Remote Routing (TRR). TLR is a mechanism that allows to reaching destinations in the vicinity of a terminode and does not use location information for making packet forwarding decisions. TRR is used to send data to remote destinations and uses geographic information; it is the key element for achieving scalability and reduced dependence on intermediate systems. The major novelty is the Anchored Geodesic Packet Forwarding (AGPF) component of TRR. This is a source-path-based method designed to be robust for mobile networks: Instead of using traditional source paths, that is lists of nodes, it uses anchored paths. An anchored path is a list of fixed geographical points, called anchor. The packet loosely follows anchored path. At any point, the packet is sent in the direction of the next anchor in the anchored path by applying geodesic packet forwarding. When a terminode finds that the next anchor geographically falls within its transmission range, it deletes it from the anchored path and sends in the direction of the new next anchor. This is repeated until the packet is sent in the direction of the final destination [24].

Authors of [16] showed by means of simulations for mobile Ad Hoc networks composed of several hundreds of terminodes, that the introduction of a hierarchy can significantly improve the ratio of successfully delivered packets and the routing overhead compared to reactive Ad Hoc routing algorithms. They also demonstrated benefits of the combination of TLR and TRR over an existing protocol that uses geographical information for packet forwarding [6, 15]. However, using greedy routing in long-distance routing makes TERMINODES inherits the problems associated with it.

4.3.11 LABAR

Location Area Based Ad Hoc Routing for GPS-Scarce Wide-Area Ad Hoc Networks (LABAR) [17] is a hybrid virtual backbone and geographical location area-based Ad Hoc routing. Authors outlined that using GPS can increase the cost and power consumption of small mobile nodes. Thus, LABAR requires only a subset of nodes (called G-nodes) to know their exact location forming location areas around them. G-nodes are interconnected into a virtual backbone structure to enable efficient exchange of information for the mapping of IP addresses to locations. Nodes that are not enabled with GPS equipment are called S-nodes.

Routing in LABAR consists mainly of three steps: zone formation, virtual backbone formation, and directional routing. The first step of LABAR deals with forming the zones, i.e., making the decision on which S-nodes should belong to which G-nodes. It was assumed that all G-nodes start the zone formation algorithm at the same time to acquire S-nodes. If an S-node has already been attached to a G-node then the request message is ignored by the S-node. Upon including an S-node in a zone, it initiates the zone formation algorithm on its own to draw more S-nodes from its neighborhood into its zone. By the end of this step, all S-nodes will belong to a G-node and G-nodes will know the IDs of their zone's S-nodes. The second step is creating an easy-to-manage virtual backbone for relaying position information of nodes. G-nodes in the virtual backbone are responsible for resolving the IP addresses into geographical locations. To connect zones and get the virtual backbone to function, a G-node called the root sends connect messages to its adjacent zones. If the particular adjacent zone is not connected yet to the backbone, then it will be added to the backbone. Figure 4.8 shows an example of such a virtual backbone.

The last step is the directional routing. The source node queries the source G-node node to map the destination IP address into the geographical location area of the destination. Then the source G-node determines the vector pointing from its own location to the destination's location. The resulting vector's direction is compared to each of the adjacent zones' direction and distance to determine the neighboring zone that will be used in relaying the data to the destination. Now, the source G-node will instruct the source node on how to route the packet inside the zone to reach the next zone with the least number of hops. The node that received the packet in the neighboring zone will route the packet to the next zone by consulting its zone's G-node (which will consume time). In the case of a failure in the directional route (determined, for example, through expired hop counters), the source zone will be informed about the failure and the virtual backbone will be used to relay the packets.

Thus, LABAR is a combination of proactive and reactive protocols, since a virtual backbone structure is used to update location information between G-nodes (in a proactive manner), while user packets are relayed using directional routing toward the direction zone of the destination. One of the most important advantages of LABAR is the reduction of cost and power consumption by the relaxation of the GPS-equipment requirement in each node.

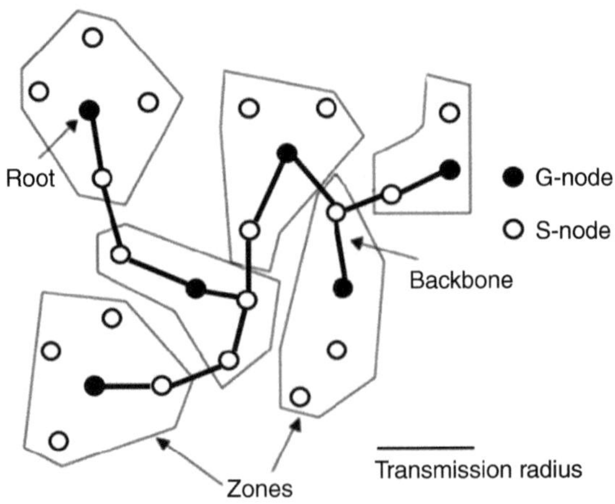

Fig. 4.8 Example of virtual backbone in LABAR

4.3.12 POSANT

Some position-based routing algorithms, such as GPSR, fail to find a route from a source to a destination (or they find a route that is much longer than the shortest path) when the network contains nodes with irregular transmission ranges. On the other hand, routing algorithms based on Ant Colony Optimization (ACO) guarantee message delivery and converge to a route which is very close to the optimum route even if the network contains nodes with different transmission ranges. However, ACO algorithms use large number of messages and need long time before the routes are established. POSition-based ANT colony routing Algorithm for mobile Ad Hoc networks (POSANT) [18] is a reactive routing algorithm which is based on ACO and uses information about the location of nodes in order to reduce the route establishment time while keeping the number of generated ants smaller in comparison to other ant-colony-based routing algorithms.

In POSANT, to establish a route from a source node S to a destination node D, neighbors of S are partitioned into three zones as shown in Fig. 4.9. After that S launches n forward ants with unique sequence numbers from each zone at regular time intervals. POSANT assumes that each node maintains a table of the values of pheromone trails assigned to its outgoing links for different destinations. Upon receiving a packet for a specific destination, a node will check if there is at least one pheromone trail for that destination, this pheromone trail will be used for making a stochastic decision to select the next hop. If no such pheromone trail exists, a pheromone trail is initialized to each outgoing link. The amount of the deposited pheromone on each link depends on the zone of the corresponding neighbor. The motivation is that

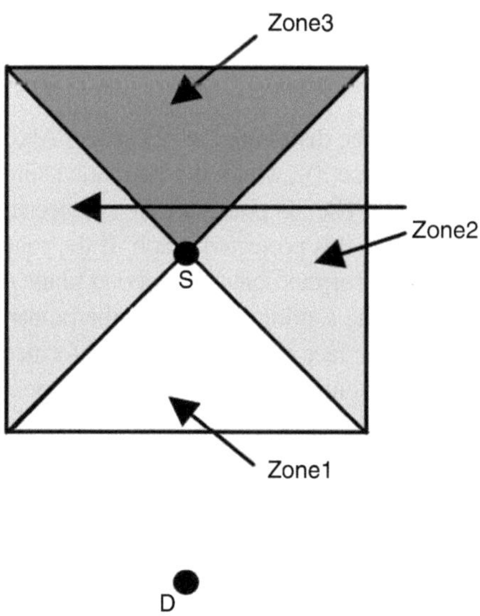

Fig. 4.9 Different zones of S for destination node D

in most cases a shortest route passes through the nodes which are closer to the direction of the destination.

Whenever a forward ant enters a node from one of its neighbors, the identifier of the neighbor, the sequence number of the ant, and the identifier of the destination will be stored. Repeated forward ants will be destroyed. When a forward ant reaches the destination, it is destroyed and a backward ant with the same sequence number is sent back to the source. Moving from node B to node A, the backward ant increases the amount of pheromone stored in edge AB. An evaporation process causes the amount of pheromone deposited in each link to decrease with time.

The above stochastic strategy establishes multiple paths between the source and destination. As a result, POSANT is a multipath routing algorithm. Multipath routing reduces the chance of congestion in the network; on the other hand, they can lead to out-of-order packet delivery problems.

If a node A realizes that the link to B is broken and there is a pheromone trail corresponding to link AB for D in the pheromone table of A. In this case the stochastic data routing will continue but if there is no pheromone trail for D in any of the other outgoing links of A, A sends a message to its neighbors to inform them that there is no route to D from A. Upon receiving this message, these neighbors do the same as if the link to A is broken. If the only outgoing link of the source node that has a pheromone trail for D breaks or a message from this link is received that states there is no route to D, a new route establishment process will begin and sending data packets will be suspended until a new route is found.

Simulations in [18] showed that POSANT has a shorter route establishment time while using a smaller number of control messages than other ant colony routing algorithms.

4.3.13 PAGs

In real applications, nodes may be distributed in 3D space. Abdallah et al. in [30] proposed the Coordinate Face (CFace(3)), which is a heuristic using a projective approach to adapt face routing [25] to 3D. The 3D points are first projected onto the xy plane and the face routing is performed on this projected graph. If the routing fails, i.e., a loop is detected, the points are then reprojected onto the second plane (the yz plane) and face routing is performed again. If the routing again fails, the points are projected onto the third plane (the xz plane) and the face routing is again performed. A simplified version of CFace(3), called CFace(1), attempts face routing with the points projected once only onto one of the xy, yz, or xz planes, randomly chosen.

In many applications, since wireless nodes are battery operated devices, they need to conserve energy so that node life is maximized. So, the same authors in [19] proposed three Power-Aware 3D Position-based Routing Algorithms for Ad Hoc Networks (called here PAGs). Authors tried to maximize the delivery rate as well as increasing network survivability (which can be measured by the remaining power in the maximum used node during a set of consecutive routing messages).

When using a fixed transmission power, nodes may waste power by transmitting with more power than is needed for correct reception. Hence, these new algorithms use the idea of replacing the constant transmission power of the node with an adjusted transmission power during neighbors' discovering and routing process stages.

The first proposed algorithm is Power Adjusted Greedy algorithm (PAG); in this algorithm all nodes use the low transmission range, which equals half of their maximum transmission range, to discover their neighbors. This process is done periodically. When a source needs to send a data packet to a specific destination the greedy routing is started. If the packet is stuck at a node that does not have a neighbor that makes progress to the destination, then the current node increases its transmission range by a specific factor and runs neighbor nodes discovery step again. If the node does not discover a new neighbor that makes progress to the destination, then the algorithm fails, otherwise greedy routing continues.

PAG has a great advantage in terms of power saving. However, through simulations authors discovered that it suffers from a low delivery rate if the network is very sparse; thus, PAG:CFace(3) was proposed. It starts with PAG routing algorithm. If the current node adjusts its transmission range and still stays in the local minimum situation, then the algorithm changes to CFace(3). If CFace(3) fails to deliver the message the algorithm fails. Their last algorithm is PAG:CFace(1):PAG. The only difference between this algorithm and the second one is that instead of trying another projective plane if the first projective plane fails, it returns immediately back to the PAG algorithm.

Simulation results showed a significant improvement in the energy saving (up to 50%) compared with available power-aware routing algorithms. Moreover, the second routing algorithm PAG:CFace(3) increases the delivery rate to around 100% for both sparse and dense networks. Hence, the most important advantage of PAGs is the great increase of the network lifetime while preserving the delivery rates. On the other hand, using greedy forwarding and discovering the nodes' neighbors using low transmission ranges will increase the probability of using paths that are longer than the optimal path. Also the adjustment of transmission range and the multiple switching among different algorithms, especially with the use of CFace(3) that requires projection of nodes into the three planes may increase processing overhead.

4.3.14 SPAAR

All the previously mentioned position-based routing protocols are vulnerable to some attacks, as they focus on improving performance while disregarding security issues [20]. In addition, most of them are not guaranteed to find the shortest path. In the past few years, a limited work has been done to introduce some security issues to position-based routing protocols. Examples of these are Secure Position Aided Ad Hoc Routing

(SPAAR) [5], Anonymous On-Demand Position-based Routing in Mobile Ad Hoc Networks (AODPR) [20] and Secure Geographic Forwarding (SGF) [21].

SPAAR uses position information in order to improve the efficiency and security of mobile Ad Hoc networks. It was designed for protecting position information in managed-hostile environment where security is a primary concern and it uses geographical information to make forwarding decisions, resulting in a significant reduction in the number of routing messages. It uses asymmetric cryptography to protect against malicious nodes (unauthorized nodes that attempt to disrupt the network) and attempts to minimize the potential for damage of attacks from compromised nodes (authorized nodes those have been overtaken by an adversary). When a node sends a multi-hop message, like a route request or a route reply, this message must be signed with its private key and encrypted with the public key of a neighbor. Every node can verify that the message was sent by a one-hop neighbor, and the destination can also verify that the sender is who it claims to be.

SPAAR achieves a high level of security by allowing nodes to only accept routing messages from one-hop neighbors. This is done to prevent the invisible node attack and the wormhole attack. To participate in SPAAR, each node requires a public/private key pair, a certificate binding its identity to its public key (signed by a trusted certificate server), and the public key of the trusted certificate server. Each node periodically broadcasts a "table update" message to inform the neighbors of its new position coordinates and transmission range. Each node maintains a neighbor table that contains the identity and position information of each verified neighbor, along with the cryptographic keys required for secure communication with each neighbor; the used location service is all-for-some.

In addition to the neighbor table, each node maintains another one for the recent destinations it has communicated with. The tables are very similar, except that the destination table also contains information about the speed of the node, making it possible to predict the next position of the node. If this is the source node's first attempt at communication with a particular destination, the source may not have the destination's position. In this situation, a location service may be used. If no location service is available, a selective flooding algorithm may be used to reach the destination and receive its position information.

To find a route to a specific destination, the source broadcasts a Route REQuest (RREQ) encrypted with its group encryption key. An intermediate node checks to see if it, or any of its neighbors, is closer to destination and it forwards the RREQ, else the RREQ is dropped. Intermediate nodes record in their route cache the address of the neighbor from which they received the RREQ, thereby establishing a reverse path. This process is repeated until the destination is reached. Upon receiving an RREQ, the destination constructs a Route REPly (RREP) signed with its private key and encrypted with the public key of the neighbor it received the RREQ from. The RREP propagates along the reverse path of the RREQ, being verified at each hop.

The fact that SPAAR makes use of geographic routing helps in reducing the overall overhead. It is also very efficient when talking about security issues; however, it requires double processing time, since it uses asymmetric cryptography, not only for end-to-end communication, but also for hop-to-hop communications [31]. SPAAR has a centralized trust and so suffers from the compromised server problem and the single point of failure.

4.3.15 AODPR

Mobile Ad Hoc networks are susceptible to malicious traffic analysis and many attacks due to the infrastructure-less, dynamic, and broadcast nature of radio transmissions. One of these attacks is the target-oriented attack which is determining a target node and conducting an intensive attack against it. In [20] Anonymous On-Demand Position-based Routing in Mobile Ad Hoc Networks (AODPR) was proposed while keeping these two problems in mind. It keeps routing nodes anonymous, thereby preventing possible traffic analysis. A time-variant Temporary IDentifier (Temp ID) is computed from time and position of a node and used for keeping the node anonymous. Moreover, AODPR uses the concept of Virtual Home Regions (VHR) which is a geographical region around a fixed center. In this scheme each node stays in one of the VHRs and nodes within a VHR obtain their own geographic position through GPS and report their position information to the Position Servers (PS). PSs are trusted Ad Hoc nodes distributed in the network. The PS keeps the position information of the nodes securely. When a node joins the network, it makes its registration to the PS and gets a common key and a pair of public and private keys from the PS [20].

When a node wants to get position information of other nodes, it first authenticates itself to the PS and sends a signed position request, and then PS provides it with the required position information, Public Key of the destination and other needed information. The source, before sending the route request, estimates Temp NH, which is initially the minimum number of hops which the route request packet travels to find a route from the source to the destination.

Each intermediate node (Forwarder) updates Temp NH, Temp NH = Temp NH - 1, and compares the updated Temp NH with the minimum number of hops which route request packet travels to find a route from this node to the destination (h′). If h′ is less than or equal to Temp NH, then forwarder forwards the packet to its radio region and keeps the route information, else it discards the packets. Both h′ and NH are calculated depending on the distance from the node to the destination and the radius of the maximum radio range coverage of each node. At the last phase of the route discovery procedure, the destination replies with a route-reply message, signed with its private key and sends this message on the reverse path.

Analysis in [20] shows that AODPR ensures the anonymity of route as well as nodes, the robustness against the target-oriented attack and several others, and it is applicable to any node density in a Network [20]. However, many fields such as NH and destination position taken from PSs are encrypted using the Common key (CK); if this key is compromised a large percentage of the communication in the whole network will be compromised. Moreover, AODPR suffers from two problems inherited from the VHR approach it uses. First, nodes can be hashed to a distant VHR, leading to increased communication and time complexity, as well as problems if the VHR of a node cannot be reached. Second, since an Ad Hoc network is dynamic, it might be difficult to guarantee that at least one position server will be present in a given Ad Hoc network [6].

4.3.16 SGF

In [21] Secure Geographic Forwarding (SGF) mechanism was proposed. It provides source authentication, neighbor authentication, and message integrity. SGF mechanism incorporates both the Hashed Message Authentication Code (MAC) [32] and the Timed Efficient Stream Loss-tolerant Authentication (TESLA) [33] with Instant Key disclosure (TIK) [34] protocol. The MAC is computed over the non-mutable part (e.g., location information of a destination) of unicast messages with the pair-wise shared secret key between the source and destination. Moreover, they proposed the usage of a reputation system, LRS, to detect and isolate message tampering and dropping attackers instead of signing the non-mutable parts of all data and control messages which may introduce too much overhead. In LRS, each node only needs to manage the reputation information of its local neighbors and periodically sends the reputation information report to its neighbors by using the HELLO messages. The TIK protocol with tight time synchronization is used to authenticate a previous forwarding node to prevent malicious users from joining a path and to avoid a message replay attack. Finally, when the destination receives a message, it can verify the authenticity of the message by comparing the received MAC to the MAC value that is computed over the received message with the secret key it shares with the source node.

In combination with SGF, a Secure Grid Location Service (SGLS) was proposed by combining SGF with the Grid Location Service (GLS) [35] so that any receiver can verify the correctness of location messages. The general concept of the proposed SGF can generally be applied to any unicast message of GLS such as location query and location reply. So the one-hop neighbor's location information can be verified by using a location verification technique [36], and the TIK protocol can be used for neighbor authentication. TESLA broadcast authentication method is used to verify the location information of two-hop neighboring nodes.

Although several forwarding strategies exist, they all forward a given message to only one optimal neighboring node based on their optimization criterion. Therefore, SGF can be applied to any of these forwarding schemes without any modification.

Simulation results in [21] showed that SGLS can operate efficiently by using effective cryptographic mechanisms. Results also showed that LRS effectively detects and isolates message-dropping attackers from the network. On the other hand, their simulations showed that the average end-to-end delay for SGLS is slightly higher than that of GLS, and that SGLS's routing overhead is much higher than that of GLS in terms of bytes which in some cases is more than double. That is, of course, due to the increase in size of routing control messages with digital signatures and MACs in SGLS.

Generally, systems using reputation system along with the a cryptography scheme in order to defend against both compromised and malicious nodes do not scale well since they have to track the reputation of all nodes, which might require huge tables of information that are difficult to manage and to keep up-to-date [31]. Moreover, SGF assumes the existence of pair-wise shared secret keys between the nodes, which is difficult to implement in large area networks. Another drawback is that SGF assumes all nodes have tightly synchronized clocks which is somewhat impractical for Ad Hoc networks. Finally, it uses the greedy forwarding which is not guaranteed to find the optimal path.

4.4 Comparison of Selected Protocols

Table 4.1 summarizes the discussed protocols together with the evaluation criteria used. This summary is a high-level qualitative comparison of the protocols rather than a precise quantitative performance evaluation. The following is an explanation of the criteria used for comparison:

- Location service type: indicates the type of the location service used with the given protocol, i.e., shows how many nodes participate in providing location information and for how many other nodes each of these nodes maintains location information.
- Location service robustness: it is considered to be low, medium, or high depending on whether the position of a given node will be inaccessible upon the failure of a single node, the failure of a small subset of the nodes or the failure of all nodes, respectively.
- Forwarding strategy type: describes the fundamental strategy used for packet forwarding.
- Forwarding strategy tolerable position inaccuracy: forwarding strategies tolerate different degrees of inaccuracy of the position of the destination. This is reflected by the tolerable position inaccuracy criterion.
- Forwarding strategy robustness: the robustness of an approach is considered to be high if the failure (or absence due to mobility) of a single intermediate node does not prevent the packet from reaching its destination. It is medium if the failure of a single intermediate node might lead to the loss of the packet but does not require the setup of a new route. Finally, the robustness is low if the failure of an individual node might result in packet loss and the setting up of a new route. According to this definition, the routing protocols that begin data transmission immediately without the need for routing setup have at least medium robustness.

Table 4.1 Characteristics of the presented forwarding strategies

Metric	Ls type	Ls robustness	Fs type	Fs tolerable position inaccuracy	Fs robustness	Fs implement. Complexity	Fs scalability	Fs packet overhead	Fs processing overhead	Loop freedom	Optimal path	Density
MFR	All-for-Some	Medium	Greedy (progress)	TR	Medium	Low	High	Low	Low	Yes [23]	Medium	Dense
DIR	All-for-Some	Medium	Greedy (direction)	TR	Medium	Low	High	Low	Low	No [23]	Medium	Dense
GPSR	All-for-Some	Medium	Greedy + perimeter	TR	Medium	Medium	High	Low	Low	Yes [24]	Medium	Both
ARP	All-for-Some	Medium	Greedy (distance + angle)	TR	Medium	Low	High	Low	Low	No (only local ones)	Medium	Both
I-PBBLR	–	–	Greedy (progress + direction)	No beacons	Medium	Low	High	Low	Low	Yes [4]	Medium	Dense
DREAM	All-for-All	High	RDF	Expected Region	High	Low	Medium	Medium	Low	No [23]	High	Both
LAR	–	–	RDF	Expected Region	Low	Low	Medium	Medium	Low	No [23]	High	Both
LARWB	–	–	RDF	Expected Region	Medium	Low	Medium	Medium	Low	No	High	Both
MLAR	–	–	RDF	Expected Region	Medium	Low	Medium	Medium	Low	Yes [14]	Low	Both
Grid	–	–	Hierarchical	Grid-by-Grid routing	Medium	Medium	High	Low	Low	Yes	Low	Dense
TERMIN-ODES	All-for-Some	Medium	Hierarchical	Short-Distance Routing Range	Medium	Medium	High	Low	Low	Yes [24]	Medium	Dense

LABAR	Some-for-All	Medium	Hierarchical	Zones	High	Medium	High	Low	Low	No	Low	Dense
POSANT	–	–	Multiple Greedy (pheromone)	Zones and pheromone	Medium	Medium	Medium	Medium	Low	Yes [18]	High	Both
PAGs	All-for-Some	Medium	PAG + CFace	TR	Medium	Medium	High	Low	Medium	Yes [19]	Low	Both
SPAAR	All-for-Some	Medium	RDF	Expected Region	Low	High	Medium	High	High	Yes	High	Both
AODPR	Some-for-Some	Medium	RDF	Expected Region	Low	Medium	Medium	Medium	Medium	Yes	High	Both
SGF	All-for-Some	Medium	Greedy	TR and neighbors' TR	Medium	High	Medium	High	High	Depends on optimization criteria	Medium	Moderate

Abbreviations: LS – location service; FS – forwarding strategy; TR – transmission range; RDF – restricted directional flooding

- Forwarding strategy implementation complexity: describes how complex it is to implement and test a given forwarding strategy. This measure is highly subjective and we will explain our opinion while discussing each protocol.
- Forwarding strategy scalability: describes the performance of the protocol with increasing number of nodes in the network. It can be classified as follows: high scalability is used when a network grows as much as it needs and the approach is still able to maintain a good performance. Medium scalability means that an approach can handle networks with a reasonable size, but may have problems if it grows. Low scalability describes protocols which restrict to small networks. Since all the position-based routing protocols are scalable, all the discussed protocols have at least medium scalability.
- Forwarding strategy packet overhead: refers to bandwidth consumption due to larger packets and/or higher number of signaling packets. The protocols can be classified as follows: Low overhead is used to describe protocols which have small packets and reduce the number of packets sent using unicast for example. Medium overhead is used to classify the protocols that have small packets but require large number of signaling packets, or if they require larger packets but use unicast to send the data. High overhead means that an approach requires larger packets as well as an increased number of signaling packets. Note that all the position-based routing protocols have lower packet overhead compared to other types, but this criterion is defined to compare the position-based routing protocols together.
- Forwarding strategy processing overhead: is used to associate each protocol with processing requirements. Low processing refers to approaches that require a low CPU processing, such as unsecure protocols. Medium processing will be used to classify the secured protocols. High processing is used to describe protocols that use multiple security techniques together.
- Loop-freedom: any routing protocol should be inherently loop-free to preserve the network resources and guarantee the correct operation of the protocol. Therefore, the discussed protocols are classified as having or not having loop-freedom property.
- Optimal path: is used to indicate the protocol probability of finding and using the shortest path for data packet relay.
- Density: indicates whether the protocol is more suitable to be implemented in a dense network, sparse network, or both.

As other greedy forwarding protocols, all nodes in MFR maintain a one-hop neighbor table, i.e., MFR uses all-for-some location service. Hence, a given node will be inaccessible upon the failure of a subset of the nodes, i.e., its location service has a medium robustness. However, the technique that is used to enable the source to know the position of the destination is not discussed. Greedy forwarding is both efficient and very well suited for use in Ad Hoc networks with a highly dynamic topology [6]. However, one important drawback of the current greedy approaches is that the position of the destination needs to be known with an accuracy of a one-hop transmission range, otherwise the packets cannot be delivered [6].

MFR robustness is medium since the failure of an individual node may cause the loss of a packet in transit, but it does not require setting up a new route, as would be the case in topology-based Ad Hoc routing. Such an approach is very easy to implement and scalable since it does not need routing discovery and maintenance [22]. Moreover, it has a low packet and processing overhead because of its fewer number of small-size packets compared to other secure-position-based protocols. MFR is probed to be a loop-free algorithm [23] since it always forces message to make a step closer to the destination. Generally, greedy routing may not always find the optimum route, even it may fail to find a path between source and destination when one exists [6, 21]; their probability of finding optimal path is considered as medium. Finally, all basic distance, progress, and direction-based methods such as MFR and DIR have high delivery rates for dense graphs, and low delivery rates for sparse graph [24].

As a greedy forwarding protocol, DIR has the same criteria as MFR, except that DIR and any other method that includes forwarding message to neighbor with closest direction, such as DREAM, are not loop-free as shown in [23].

GPSR also exhibits all the properties of greedy forwarding except that its implementation effort is considered to be of medium complexity due to planarizing underlying network and using perimeter routing. On the other hand, using the right hand thumb rule and perimeter mode routing made it applicable in sparse networks as well as dense ones.

ARP exhibits all the properties of greedy forwarding except that memorizing the last visited hops in the packet header avoids local loops, but does not guarantee its loop freedom. Moreover, using angle-based forwarding scheme to circumvent voids makes it applicable in sparse networks as well as dense ones.

I-PBBLR inherited all the properties of greedy forwarding; however, the used location service was not discussed at all. Moreover, using a beaconless protocol slightly increases the robustness and scalability, reduces the packet overhead, improves the performance in sparse networks, and increases tolerability to position inaccuracy compared to traditional greedy protocols. Finally, using the improved progress, guarantees loop freedom as the packets are always forwarded a step toward the destination.

Restricted directional flooding protocols, such as DREAM and LAR, are robust against position inaccuracy since they use the expected region concept. They have higher communication complexity than greedy ones and, therefore, have less scalability to large networks; their scalability and packet overhead are considered to be medium. However, their processing overhead is low due to nonsecure routing. Moreover, both of them include forwarding packets to neighbors with closest direction, so both are not loop-free [23]. On the other hand, they are very simple to implement and have high probability of finding the optimal path. Finally, they may be implemented in both dense and sparse networks; they are better for sparse networks than greedy forwarding, and even if they are used in dense ones they will have good performance due to low processing and medium packet overheads.

DREAM's location service is fundamentally different from other location services in that it requires that all nodes maintain position information about every other node. This leads to large overhead due to position update and large position information maintained by each node. Hence, DREAM's location service is the least scalable position service and, thus, not appropriate for large-scale and general-purpose Ad Hoc networks. On the other hand, a position query requires only a local lookup and the position of a given node will be inaccessible only upon the failure of all nodes, which makes it very robust. LAR however, does not use such location service; it just uses the available position information from a route that was established earlier.

DREAM is very robust against the failure of individual nodes since the data packet goes through multiple paths, so the failure of a single intermediate node does not prevent the packet from reaching its destination. This qualifies it for applications that require a high reliability and fast message delivery for very infrequent data transmissions [6]. LAR is robust during the route discovery since the RDP packet goes through multiple paths; however, after route setup it is like any other protocol that depends on route setup before sending the data packets, i.e., the failure of a single node might result in packet loss and the setting up of a new route. Hence, its robustness is considered to be low. On the other hand establishing a route before beginning data sending makes it more suitable than DREAM in the cases that require high volume of data transmissions.

LARWB exhibits all the properties of LAR except that its robustness is considered to be medium since the failure of a single node might result in packet loss but does not result in setting up of a new route due to the usage of the route backup. This route backup also reduces the number of routing packets; however, we still cannot consider LARWB's packet overhead as low as that in greedy.

MLAR has similar criteria as that of LAR except that its robustness is medium since the failure of a single node might result in packet loss but does not result in setting up of a new route due to the usage of the alternate paths. These alternate paths also reduce the packet overhead; however, it is still higher than that of greedy. Since MLAR caches the most recently received routes, the probability of using the optimal path is very low. Lastly, since the packet header contains the entire source route, all paths are checked easily as being loop-free at each node that stores routes; loop freedom is guaranteed in MLAR.

Grid is a hierarchical routing that applies the concept of dominating sets. It is like LAR, uses the available position information of the destination from a route that was established earlier to implement a restricted directional flooding among grids. So it is robust against position inaccuracy since it uses Grid-by-Grid routing and expected region concept. Although Grid has strong route maintenance capability and it is very robust toward node mobility, it is like any other protocol that depends on route setup before sending the data packets in the sense that the failure of a single node might result in packet loss and the setting up of a new route. Moreover, the authors in [15] did not elaborate on route maintenance required when a grid remains empty after its leader and only node leaves it [24]. Thus, its robustness is considered to be medium.

Grid's implementation complexity is considered to be medium due to dealing with the area as grids. Its scalability is high due to using restricted directional flooding and delegating the searching responsibility to the gateway hosts. Its packet and processing overheads are considered to be low due to reduced number of small nonsecure routing packets. Grid uses a specific field to detect duplicate request packets from the same source, so endless flooding of the same request can be avoided, i.e., it is loop-free. On the other hand, GRID uses long paths since packets are forced to be routed through grid leaders.

Finally, it is better to implement Grid in dense networks due to performing routing in a grid-by-grid manner through grid leaders. So, the number of packets related to route search is insensitive to the network density. On the contrary, the cost slightly goes down as the host density increases, since routes become more stable with denser hosts. On the other hand, if it is implemented in sparse networks, each node will be the gateway of its grid, and GRID may become like native LAR; so it will consume network resources in dividing the area into grids and electing gateways without any benefit.

TERMINODES provide hierarchical approach to position-based Ad Hoc routing. For long-distance routing it uses a greedy approach and, therefore, has characteristics similar to those of greedy forwarding. However, due to the usage of a non-position-based approach at the local level, it is more tolerant to position inaccuracy. As other greedy forwarding protocols, in TERMINODES all nodes maintain a one-hop neighbor table; it uses all-for-some location service. Hence, a given node will be inaccessible upon the failure of a subset of the nodes; its location service has a medium robustness. Moreover, it may fail to find the optimum route and has higher delivery rates for dense graphs.

TERMINODES robustness is medium since the failure of an individual node may cause the loss of a packet in transit, but it does not require setting up a new route, as would be the case in topology-based Ad Hoc routing. Due to using the two-level hierarchy approach, TERMINODES is considered to have medium implementation complexity. Such an approach is scalable since it does not need routing discovery and maintenance in long-distance routing. Moreover, it has a low packet and processing overhead because of its fewer number of small-size packets compared to other secure-position-based protocols. TERMINODES is considered to be a loop-free algorithm [24] since it always forces message a step closer to the destination.

LABAR is a hierarchal protocol since it uses zone-based routing. In LABAR the virtual backbone structure is used to update location information between Gnodes in a proactive manner; the used location service type is some-for-all. Generally, the robustness of such approaches is medium, since the position of a node will become unavailable if a subset of the nodes fails. LABAR exhibits some properties of greedy forwarding such as high scalability, low packet overhead, and its suitability to be implemented in dense networks. LABAR tolerates position inaccuracy due to relaying the user packets toward the direction of the destination's zone and not toward the exact position of it.

In the case of a failure in the directional route of LABAR, the virtual backbone will be used to relay the packets, i.e., LABAR's robustness is high since a failure of a single intermediate node does not prevent the packet from reaching its destination. LABAR's implementation complexity is considered to be medium due to using zones and its processing overhead is low due to nonsecure routing. LABAR is not considered a loop-free protocol since it uses directional flooding and does not use any technique to indicate that a specific packet has been received by a specific node before this time. However, the expired hop counters can be used to stop the loops after a while.

One may think that LABAR's probability of finding optimal path is medium since it uses greedy routing; however, their simulations showed that LABAR usually uses a long path which sometimes reaches more than double the length of the optimal path. This may be due to using directional route toward the direction of the destination's zone and not toward the exact position of it. So, LABAR's probability of findinf optimal path is considered to be low.

In POSANT, the used location service was not discussed. However, the used forwarding strategy is multiple greedy forwarding with the pheromone trail value used as the optimization criteria; a source launches many forward ants from different zones at regular time intervals and the pheromone trails values are used for making a stochastic decision to select the next hop.

POSANT tolerates position inaccuracy by forwarding ants to different zones (not to specific node position) and depending on the pheromone trail value as the optimization criteria (which does not depend on the exact position of nodes). POSANT's robustness is considered to be medium, since the failure of a single node might result in packet loss but does not result in a new route establishment, except if the only outgoing link of the source node that has a pheromone trail for D breaks or a message from this link is received stating that there is no route to D.

Using periodic multiple greedy forwarding made POSANT's implementation complexity, scalability, and packet overhead considered as medium. However, its processing overhead is low due to nonsecure routing. POSANT is guaranteed to be loop-free since repeated forward ants are destroyed. Moreover, it has a high probability of finding the optimal path since it is based on ACO which guarantees message delivery and converges to a route which is very close to the optimum route even if the network contains nodes with different transmission ranges.

Finally, POSANT may be implemented in both dense and sparse networks. POSANT is better for sparse networks than traditional greedy forwarding; since if no pheromone trail exists, the route discovery packet will not be dropped; however, pheromone trail initialization is done. Moreover, if it is used in dense networks it will have good performance due to low processing and medium packet overheads.

In the PAGs, all nodes use the low transmission range to discover their neighbors periodically; the used location service is all-for-some. Thus, a given node will be inaccessible upon the failure of a subset of the nodes; its location service has a medium robustness. However, the technique that is used to enable the source to know the position of the destination is not discussed.

The used forwarding strategy is a combination of power-adjusted greedy and CFace algorithms. As with other greedy forwarding protocols, the position of the destination needs to be known with an accuracy of a one-hop transmission range, otherwise the packets cannot be delivered [6]. Moreover, PAGs have high scalability, low packet overhead, and low probability of finding the shortest path. The robustness of PAGs is medium since the failure of an individual node may cause the loss of a packet in transit, but it does not require setting up a new route. Their implementation complexity is also considered as medium due to the usage of both greedy and CFace algorithms.

Pure greedy forwarding algorithms have low processing overhead; however, PAGs' processing overhead is considered to be medium due to the adjustment of transmission range and the multiple switching among different algorithms, especially with the use of CFace(3) that requires projection of nodes into the three planes. CFace detects the loops, i.e., PAGs are loop-free algorithms. PAGs' probability of finding an optimal path is low due to using greedy forwarding and discovering the nodes' neighbors, using low-transmission ranges. Finally, PAGs can be implemented in both sparse and dense networks (considering the delivery rate and network lifetime). Their simulations support especially the second routing algorithm PAG:CFace(3), which increased the delivery rate to around 100% for both sparse and dense networks. The advantage that can be obtained when the PAGs are implemented in dense networks is the high probability of finding a path even with nodes working at low transmission level; no need to change among different algorithms. However, this will lead to longer paths. On the other hand, implementing PAGs in sparse networks will increase the probability of finding a path compared to pure greedy algorithms, but time will be lost in switching to high transmission range and other CFace algorithms.

In SPAAR, each node maintains a neighbor table that contains the identity and position information of each verified neighbor; the used location service type is all-for-some. The source node can calculate the approximate geographic location of the destination from the most recent location and most recent velocity information stored in the source node's destination table. On the first attempt at communication with a particular destination, the source may use a location service or a selective flooding algorithm to reach the destination and receive its position information. The general robustness of this approach is medium, since the position of a node will become unavailable if a significant number of nodes fail.

SPAAR uses the restricted directional flooding, so it exhibits some properties such as the high probability of using the optimal path. Moreover, it is loop-free since it depends on forwarding the packet to the nodes toward the destination and uses sequence number.

SPAAR tolerates position inaccuracy by the expected region; each node forwards the RREQ only if it, or any of its neighbors, is closer to the destination. Its robustness is low since the failure of an individual node might result in packet loss and the setting up of a new route. SPAAR has high implementation complexity since messages must be verified, signed with the private key and encrypted with the public key of a neighbor. But it is still less than the complexity of SGF since there is no reputation system.

SPAAR assumes the existence of one certificate server, which may be the operation bottleneck especially in large area networks. Moreover, increasing the number of nodes in the network with using the restricted directional flooding will increase the packet overhead. Finally, in large area networks the probability of having long routes will increase, and since each node spends time in signing and encrypting the messages, the probability of node movements and route breakage will increase. For these three reasons SPAAR is considered to have a medium scalability. Moreover, SPAAR has a high packet overhead because of the large-size packets due to the security techniques used and increased number of packets compared to greedy forwarding. These security techniques lead also to high processing overhead.

Finally, SPAAR can be implemented in both, sparse and dense networks. It is suitable for sparse networks since it uses the restricted directional flooding, not greedy. At the same time, it is suitable for dense networks since increasing the number of neighbors will case larger neighbor table but the computational overhead for the encryption of messages remains constant [5].

In AODPR each PS keeps the position information of the nodes that hashed into its VHR; hence, the used location service type is some-for-some. Accordingly, a given node will be inaccessible upon the failure of the PSs of its VHR, i.e., its location service has a medium robustness. AODPR uses the restricted directional flooding, so its probability of using the optimal path is high. Moreover it is loop-free since it depends on forwarding the packets to the nodes toward the destination and uses sequ ence number.

AODPR tolerates position inaccuracy by using the expected region. Its robustness is low since the failure of an individual node might result in packet loss and the setting up of a new route. AODPR's implementation complexity is considered to be medium since messages are signed only with the private key of each node. So its complexity is less than SPAAR and SGF, since it does not use neighbor public key or reputation system.

AODPR has a medium scalability, since increasing the number of nodes in the network with the usage of the restricted directional flooding will increase the packet overhead. However, it still has a higher scalability than SPAAR due to the reasons mentioned in the discussion of SRAAR scalability. AODPR also has a less packet overhead compared to SRAAR. Even though the number of sent packets in AODPR is large, its packet size is smaller than that in SPAAR due to the later security techniques; AODPR is considered to have a medium packet overhead and processing overhead.

Finally, AODPR is applicable to any node density in a network [20]. It is suitable for sparse networks since it uses the restricted directional flooding, not greedy. At the same time, it is suitable for dense networks, since increasing the number of nodes will cause larger position information tables in the PSs without affecting the computational overhead for the encryption of messages.

In SGF each node should maintain a table of its immediate neighbors as well as each neighbor's neighbors [21]. So the used location service type is all-for-some. Accordingly, a given node will be inaccessible upon the failure of a subset of the nodes; the robustness of its location service is medium. SGF uses the greedy forwarding, so it exhibits some greedy properties such as uncertainty to use the optimal path and the medium robustness. SGF tolerates position inaccuracy by the list of neighbors HELLO messages that each node periodically broadcasts; each node knows the exact position of the nodes in its transmission range and neighbors' transmission ranges.

It is clear that it is so complex to implement SGF since it uses many securing techniques whether with the location service or the forwarding strategy. SGF assumes the existence of pair-wise shared secret keys between the nodes which is difficult to implement in large area networks; so it has medium scalability. Moreover, it has a high packet overhead due to the periodically sent reputation information report and list of neighbors HELLO messages, in addition to the large-size packets due to the security techniques used. These security techniques lead also to high processing overhead. SGF loop freedom depends on the used optimization criteria (directional or other). Finally, SGF is preferred to be implemented in moderate density networks; since greedy forwarding may have problems in sparse networks. On the other hand, implementing it in a dense network will increase the size of the periodic list of neighbors and reputation information HELLO messages, which may consume the network bandwidth and nodes' memory.

4.5 Directions of Future Research

In this chapter we have shown that there are many approaches to perform position-based packet forwarding. However, there still exist a number of issues and problems that need to be addressed in future research.

Position-based protocols make it possible to have larger networks without scalability problems. However, geographical routing also offers attackers new opportunities, especially since most protocols broadcast position information in the clear allowing anyone within range to receive. Hence, node position can be altered, making other nodes believe that it is in a different position. This may make nodes believe that the attacker is the closest node to the destination and choose it as the next hop. Consequently, this attacker will be able to alter or drop packets. Thus, it is of worth that more intensive work be done to enable secure-position-based routing protocols to defend against several attacks not only from malicious nodes, but also from the compromised ones. Additionally, location privacy is one of the most major issues that need to be addressed, since location privacy is hard to achieve when a node identifier can be immediately associated with its position. Finally, alternative security schemes that are not based on infrastructure for key distribution should be considered, especially since they suffer from high processing requirement (due to signing and signature verification of every packet) and may be a perfect target for Denial of Service (DoS) where attackers try to exhaust a nodes processing time and battery, by forcing them to spend time doing cryptographic calculations that are not required. Moreover, approaches that suggest the usage of symmetric cryptography suffer from a scalability problem since every pair of nodes would require a unique shared key.

Geographical routing protocols depend strongly on the existence of distributed scalable location services, which are able to provide the location of any host at any time throughout the entire network. Hence, researches should consider the scalability point upon developing new location services. Also, the most common way to enable nodes to know their locations is by equipping them with GPS. To decrease the cost and power consumption of small mobile nodes other techniques for finding relative coordinates should be discussed.

We also need more concentration on power-aware routing for saving network energy by developing protocols that have as many as possible sleeping nodes and designing sleep period schedules for each node. Also, more studies should concentrate on Quality of Service (QoS) position-based routing and multicast position-based routing.

Most routing protocols (not only position-based) consider nodes as neighbors if the Euclidean distance between them is at most equals the transmission radius which is the same for all nodes in the network. However, irregular transmission radius of a node (due to obstacles or noise), unidirectional links, and different nodes' transmission radii should be taken into consideration. Moreover, many applications have nodes distributed in three-dimensional space and few researches have been done in this field as yet.

Another issue that needs to be addressed is enabling the connectivity among the individual Ad Hoc networks, as well as connectivity of any given Ad Hoc network to the Internet. This will, most likely, require the usage of hierarchal approaches to achieve scalability. This field has already begun; however, it needs further investigation.

4.6 Summary and Conclusions

Efficient routing among a set of mobile hosts is one of the most important functions in Ad Hoc wireless networks. Many points should be taken into consideration when developing a routing protocol; some of them are high delivery rate, short hop count, small flooding ratio, small end-to-end delay, low power consumption, high level of security, and avoiding single point of failure.

This survey has presented the current state of position-based Ad Hoc routing and provided a qualitative evaluation of the presented approaches. At the end, we identified a number of research opportunities which could lead to further improvements in position-based Ad Hoc routing.

Forwarding techniques based on position information was classified into three distinct categories. Greedy routing does not require the maintenance of explicit routes; however, it works by forwarding a single copy of data packet toward the destination. If a local maximum is encountered, a repair strategy can be used to avoid dropping the packet. After the comparison of the existing solutions we can conclude that the greedy packet forwarding is an efficient approach that scales well even with highly dynamic networks, and it is a promising strategy for general-purpose position-based routing. However, it is not guaranteed to find the optimal path or it may not find a path at all.

In restricted directional flooding the packets are broadcasted in the general direction of the destination. On their way, the position information in the packets may be updated if a node has more current information about the destination's position. Restricted directional flooding has higher packet overhead and less scalability. However, its opportunity to find the shortest path is higher. Using restricted directional flooding to set up a route in an efficient manner (such as in LAR) increases the probability of finding the optimal path and is suitable for the cases that require high volume of data transmissions. However, when it is used to forward the data packets themselves (such as in DREAM) it will be more suitable for situations where a small number of packets needs to be transmitted very reliably.

Using hierarchical approaches increases the approach scalability. This may be done through the usage of zone-based routing, dominating sets, or by means of a position-independent protocol at the local level and a greedy variant at the long-distance level.

Security has recently gained a lot of attention in topology-based routing protocols and many attempts in proposing end-to-end security schemes have been done. However, it is obvious from the analysis that few research efforts have addressed position-based security issues. Moreover, using multiple paths between source and destination helps in increasing the robustness of transmitted data confidentiality. However, multipath routing may cause more collision among different routes which in turn degrades network performance such as packet delivery ratio. Finally, a few researchers have considered the power efficiency metric while developing their protocols.

The main prerequisite for position-based routing is that a sender can obtain the current position of the destination. Therefore, our next step is making a survey about existing location services.

References

1. Giruka V, Singhal M (2005) Angular routing protocol for mobile ad-hoc networks. In: Procee dings of the 25th IEEE international conference on distributed computing systems workshops (ICDCSW'05). Kentucky University, Lexington, KY, pp 551–557
2. Perkins C, Royer E (1999) Ad hoc on-demand distance vector routing. IEEE proceedings of the 2nd IEEE workshop on mobile computing systems and applications, New Orleans, LA, pp 90–100
3. Abolhasan M, Wysocki T, Dutkiewicz E (2004) A review of routing protocols for mobile ad hoc networks. Ad hoc Netw 2(1):1–22, Elsevier
4. Cao Y, Xie S (2005) A position based beaconless routing algorithm for mobile ad hoc networks. Proc Int Conf Commun Circuits Syst 1(1):303–307, IEEE
5. Carter S, Yasinsac A (2002) Secure position aided ad hoc routing. Proceedings of the IASTED international conference on communications and computer networks (CCN02), Cambridge, pp 329–334
6. Mauve M, Widmer J, Hartenstein H (2001) A survey on position-based routing in mobile ad-hoc networks. IEEE Netw 15(6):30–39
7. Qabajeh LK, Kiah M, Qabajeh M (2009) A qualitative comparison of position-based routing protocols for ad-hoc networks. IJCSNS 9(2):131–140

8. Takagi H, Kleinrock L (1984) Optimal Transmission ranges for randomly distributed packet radio terminals. IEEE Trans Commun 32(3):246–257
9. Kranakis E, Singh H, Urrutia J (1999) Compass routing on geometric networks. Proceedings of the 11th Canadian conference on computational geometry, Vancouver, pp 51–54
10. Karp B, Kung H (2000) GPSR: greedy perimeter stateless routing for wireless networks. Proceedings of the 6th annual ACM/IEEE international conference on mobile computing and networking (MOBICOM 2000), Boston, MA, pp 243–254
11. Basagni S, Chlamtac I, Syrotiuk V, Woodward B (1998) A distance routing effect algorithm for mobility (DREAM). Proceedings of the 4th annual ACM/IEEE international conference on mobile computing and networking (MOBICOM), Dallas, TX, pp 76–84
12. Ko Y, Vaidya N (2000) Location-Aided Routing (LAR) in mobile ad hoc networks. Wireless Netw (WINET) 6(4):307–321, ACM
13. Kalhor S, Anisi M, Haghighat A (2007) A new position-based routing protocol for reducing the number of exchanged route request messages in Mobile Ad-hoc Networks. Proceedings of the second international conference on systems and networks communications (ICSNC 2007), IEEE, pp 13
14. Nanda S, Gray R (2006) Multipath Location Aided Routing in 2D and 3D. IEEE WCNC 1:311–317
15. Liao W, Tseng Y, Sheu J (2001) GRID: a fully location-aware routing protocols for mobile ad hoc networks. Telecommun Syst 18:61–84
16. Blazevic L, Buttyan L, Capkum S, Giordano S, Hubaux J, Le Boudec J (2001) Self-organization in mobile ad-hoc networks: the approach of terminodes. IEEE Commun Mag 39(6):166–174
17. Zaruba G, Chaluvadi V, Suleman A (2003) LABAR: Location Area Based Ad Hoc Routing for GPS-scarce wide-area ad hoc networks. Proceedings of the first IEEE international conference on pervasive computing and communications (PerCom'03), March 23 – 26, 2003, Fort Worth, Texas, USA, pp 509–513
18. Kamali S, Opatrny J (2007) POSANT: a position based ant colony routing algorithm for mobile ad-hoc networks. In: Proceedings of the third international conference on wireless and mobile communications (ICWMC'07). IEEE, Concordia University, Montreal
19. Abdallah A, Fevens T, Opatrny J (2007) Power-aware 3D position-based routing algorithms for ad hoc networks. Proceedings of the IEEE International Conference on Communications (ICC 2007), Glasgow, Scotland, June 24 – 28 2007, pp 3130–3135
20. Mizanur Rahman Sk, Mambo M, Inomata A, Okamoto E (200) An anonymous on-demand position-based routing in mobile ad hoc networks. In: Proceedings of the international symposium on applications and the Internet. IEEE, Mesa/Phoenix, Arizona
21. Song J, Wong V, Leung V (2007) Secure position-based routing protocol for mobile ad hoc networks. Elsevier Ad Hoc Netw J 5(1):76–86, Elsevier
22. Wu X (2005) VPDS: Virtual home region based distributed position service in mobile ad hoc networks. Proceedings of the 25th IEEE international conference on distributed computing systems (ICSCS 2005), Columbus, Ohio, USA June 06-June 10, pp 113–122
23. Stojmenovic I, Lin X (2001) Loop-free hybrid single-path/flooding routing algorithms with guaranteed delivery for wireless networks. IEEE Trans Parallel Distr Syst 12(10):1023–1032
24. Giordano S, Stojmenovic I, Blazevic L (2003) Position based routing algorithms for ad hoc networks: a taxonomy. In: Cheng X, Huang X, Du DZ (eds) Ad hoc wireless networking. Kluwer, pp 103–136 http://www.site.uottawa.ca/~ivan/routing-survey.pdf
25. Bose P, Morin P, Stojmenovic I, Urrutia J (1999) Routing with guaranteed delivery in ad hoc wireless networks. Proceedings of the 3rd international workshop on discrete algorithms and methods for mobile computing and communications,August 20 – 20, 1999, Seattle, Washington, United States
26. Johnson D, Maltz D (1996) Dynamic source routing in ad hoc wireless networks. In: Mobile computing, vol 353. Kluwer Academic, Norwell, MA, pp 153–181
27. Marina M, Das S (2001) On-demand multipath distance vector routing in ad hoc networks. Proceedings of the 9th international conference on network protocols (ICNP 01), November 11 – 14, 2001, Mission Inn, Riverside, California, pp 14–23

28. Feeney L, Nilsson M (2001) Investigating the energy consumption of a wireless network interface in an ad hoc networking environment. Proceedings of the Twentieth Annual Joint Conference of the IEEE Computer and Communications Societies (INFOCOM 2001), vol 3. Anchorage, AK, pp 1548–1557
29. Shih E, Cho S, Ickes N, Min R, Sinha A, Wang A, Chandrakasan A (2001) Physical layer driven protocol and algorithm design for energy-efficient wireless sensor networks. In: Proceedings of the seventh annual international conference on mobile computing and networking (MOBICOM 2001). ACM, Rome, Italy, pp 272–287
30. Abdallah A, Fevens T, Opatrny J (2006) Randomized 3-D position based routing algorithm for ad-hoc networks. Proceedings of the 3rd annual international conference on mobile and ubiquitous systems: networks and services (MOBIQUITOUS), San Jose, CA, pp 1–8
31. Fonseca E, Festag A (2006) A survey of existing approaches for secure ad hoc routing and their applicability to VANETS. NEC Technical Report NLE-PR-2006–19, NEC Network Laboratories. http://www.network-on-wheels.de
32. Krawczyk II, Bellarc M, Canctti R (1997) HMAC: keyedhashing for message authentication. Internet Engineering Task Force, Request for Comment RFC 2104. http://www.ietf.org/rfc/rfc2104.txt
33. Perrig A, Canetti R, Song D, Tygar D, Briscoe B (2004) TESLA: multicast source authentication transform introduction. Internet Engineering Task Force, Internet Draft of Multicast Security Working Group. http://tools.ietf.org/html/draft-ietf-msec-tesla-intro-04.txt
34. Hu Y, Perrig A, Johnson D (2003) Packet leashes: a defense against wormhole attacks in wire-less network. Proceedings of the 22nd Annual Joint Conference of the IEEE Computer and Communications Societies (INFOCOM 2003) 3(1):1976–1986, April 1 – 3, 2003 San Francisco, CA, USA
35. Li J, Jannotti J, De Couto D, Karger D, Morris R (2000) A scalable location service for geographic ad hoc routing. Proceedings of the 6th Annual ACM/IEEE International Conference on Mobile Computing and Networking (MOBICOM 2000), Boston, MA, pp 120–130
36. Capkun S, Hubaux J (2005) Secure positioning of wireless devices with application to sensor networks. Proceedings of the 24th Annual Joint Conference of the IEEE Computer and Communications Societies (INFOCOM 2005), 3(1), Miami, Fl, pp 1917–1928

Part II
Multidisciplinary Issues

Chapter 5
Sensor Networks' Integration

Szymon Fedor, Alex Gluhak, and Srdjan Krco

5.1 Introduction – Motivations for SN Integration Frameworks

5.1.1 Picture of Current WSN Deployments, Problems Related

Sensor networks and applications thereof have been intensively researched in the past decade and a variety of systems have been meanwhile deployed in real-world settings. Most of these applications and the corresponding sensor networks they use are designed as vertically integrated systems [1–3]. In such vertical systems, a sensor network or a limited set of mostly homogeneous sensor networks are deployed for a specific application in mind. The application is mostly the sole user of this sensor network and has a priori knowledge of the capabilities that the sensor network(s) provides. An application also typically knows how to address the respective gateways/sinks of the sensor networks, in order to interact with the sensor networks and shares a common interaction protocol with them.

As the number of the sensor networks that may be used by an application grows, it is becoming increasingly cumbersome for applications to manage direct interactions between those. Furthermore, the reuse of the existing sensor network infrastructure for multiple applications could avoid redundant deployment of similar sensor networks at the same location and provide higher returns for the initial investments costs of the deployed sensor network infrastructure. Recent research has therefore focused on overcoming the inflexibility of the tightly coupled vertical system and proposed several sensor network integration frameworks [3–6]. These frameworks aim to break up the vertical systems into horizontal reusable system components and make them available to a larger set of applications. The frameworks typically provide support functionality that significantly reduces

---

S. Fedor (✉), A. Gluhak, and S. Krco
Ericsson Ireland
e-mail: szymon.fedor@ericsson.com

L. Gavrilovska et al. (eds.), *Application and Multidisciplinary Aspects of Wireless Sensor Networks*, Computer Communications and Networks,
DOI 10.1007/978-1-84996-510-1_5, © Springer-Verlag London Limited 2011

the interaction complexity of applications and eases incremental deployment of new sensor networks. Via these frameworks, applications can gain access to a large variety of connected geographically distributed sensor networks.

While representing first stepping stones for a real-world Internet, a variety of different issues remain unaddressed whereas they are essential for realizing an ecosystem for real-world contexts and interactions. Therefore, the development of sensor network integration frameworks is currently being carried by many industrial and academic institutions. In this chapter, an overview of existing sensor network integration frameworks (SNIFs) is presented, highlighting the main concepts and key features. Various examples of these frameworks are provided covering different design approaches from both industrial and academic organizations. Each of these frameworks is briefly analyzed with the description of key features and innovative solutions. Also their potential limitations and shortcomings are highlighted.

5.1.2 Benefits of Integration

The sensor network integration frameworks provide various advantages for different parties: end users, service providers, WSN providers, network operators, and network service providers. This paragraph describes how the numerous features of SNIF can be beneficial for those entities.

First, better visibility of WSNs obtained with SNIF can be advantageous for the users of WSNs. They can avail of the broader range of measurements which are closer to the phenomena of interest and therefore are more precise. Besides, the availability of diverse sensors would allow deployment of a more sophisticated and complex applications. Also, greater number of available measurements will increase competition between WSNs providers and as a result will lead to a lower cost of those services. All these factors will contribute to a so-called community effect where more and more users will benefit from the WSNs. As a consequence, WSN providers would compensate more quickly the investments incurred for the deployment and maintenance of WSNs. Also the service providers and network operators would benefit from the greater visibility of the WSNs which will increase the demand for their services. In addition, the availability of a greater number of diverse sensors would allow the development of more elaborate services requiring measurements of various quantities.

Also the resource naming and discovery provided by SNIF will bring benefits to various entities. Currently, many WSNs are deployed for a specific application as vertical solutions. With SNIF, these systems could be easily discovered when needed by other applications. In consequence, the service providers can offer more complex applications on the basis of already existing infrastructures. Also it would be beneficial to the end users having access to a broader range of applications.

Most of the SNIFs provide the common interface for the WSN measurements. This will reduce significantly the effort required to access the data provided by WSNs. In consequence, the service providers could develop much quicker applications using WSNs. The current situation of the WSNs market with various independent systems makes difficult the possibility of establishing a connection to a WSN gateway. With SNIF, the access to the WSNs via a standard interface will be much easier and will require a minimal effort from the service providers.

Some of the major advantages of the SNIFs are the mechanisms which enable security of shared resources against misuse, provide estimates of reliability or verifiability of sensed data against malicious intervention or inadvertent errors, and protect the privacy of users who are being sensed or sharing parts of their data. These security issues are always a concern in shared systems and in most currently deployed WSNs they are not considered because the access to the sensor measurements is granted to a known and limited number of users. The security mechanisms are especially beneficial for the end users and WSNs providers who will be keener to share their measurements if the data are handled securely. As a result, more services using WSNs would be developed and offered, which would be advantageous for service providers and network operators.

The security enhancements provided by SNIF would enable the support of accountability, access control, and billing methods. These features would facilitate development of business models based on the WSNs and therefore more investors would be interested in investing in WSN applications.

Another functionality of SNIF which will also contribute to the expansion of the WSN market is a high-level query processing. Current WSN systems handle very basic queries and because they are designed as vertical solutions, they are not able to respond to complex queries which would require interaction with various heterogeneous WSNs. This feature would enable deployment of sophisticated, complex applications and composed services.

A different feature of SNIF which would enable new applications is caching and data history. Many end users will be interested not only in applications using current and future measurements but also in applications which require sensor measurements from the past. Most current WSN systems can respond to requests for measurement of future events. The previously obtained sensor samples can only be reused if the user explicitly requests to store them for a potential future application. With SNIF, the WSN measurements can be shared among various applications. Therefore, it is possible that the sensor measurements used for one service can be requested in the future by another application. As a result SNIFs would allow service providers to offer applications which require sensor data from the past.

5.2 Existing Integration Frameworks

5.2.1 Overview

5.2.1.1 Historic Perspective on Integration Frameworks Deployment

The research and development in WSNs was initially driven by defense applications. Around 1980 the Defense Advanced Research Projects Agency (DARPA) started the Distributed Sensor Networks (DSN) program in order to study whether the agency's approach for communication could be extended to sensor networks. This research program resulted in many WSN systems like acoustic tracking of low-flying aircraft [7] or Remote Battlefield Sensor System (REMBASS). These solutions were very expensive and could be used only for a dedicated military purpose. The main focus of the WSN research remained in the military area until the end of 1990s when the first motes for environmental monitoring were developed [8].The availability of low-cost sensor nodes has resulted in the emergence of many other potential applications, from industrial sensing to infrastructure security and health care. Since the end of 1990s many companies have created vertical WSN systems which can only be used for a single application. To stimulate WSN market development by increasing the interoperability between these dedicated systems, the companies developed communication standards for WSNs (Zigbee [9], WirelessHART [10] and ISA-SP100 [11]). Although these standards are already mature and many companies sell products complaint with them, WSN market has not expanded significantly since their publication. Therefore currently, most WSN systems are deployed as vertical solutions and the users do not tend to apply them for multiple applications. As a result the price of WSN nodes is still high and the WSN suppliers try to respond to the market demand by developing systems only for dedicated application.

The WSN research community realized the weaknesses of the vertical solutions (described in Section 5.1.1) and one of the first solutions that addresses these problems was IrisNet project from Intel Research started around 2000 [12]. The project aims to develop a scalable software infrastructure that employs data mining to let Internet users query real-time and historical video information produced by Web cams and other sensors. ÌrisNet takes database-centric approach in its design and users can query for measured data using XPATH query language (see Section 5.2.2.1).

Since the emergence of IrisNet system many solutions have been proposed which facilitate deployment of horizontal applications. They share many features of IrisNet, e.g., database-like design approach (Hourglass [13], SenseWeb [14]) or multi-tier architecture (JWebDust [15], Janus [6]). Existing sensor networks integration frameworks differ mainly by the number of provided features and also by the maturity of the system implementation. Some of them have only been implemented as a prototype (e.g., Janus) and others have been widely used for various applications and are constantly being improved by the research community (e.g., GSN). But none of the proposed solutions have been used in a commercial application and only the future can show if the sensor networks integration frameworks would contribute to the expansion of the WSN market.

5.2.1.2 Summary of Features of the Existing Integration Frameworks

Table 5.1 summarizes nonfunctional properties of the surveyed existing SNIFs. The detailed analysis of each presented SNIF is provided below. Most of the proposals enable the interactions of applications with different heterogeneous SANs and support many necessary types of interactions such as query- and event-based interactions or streaming. Some approaches such as IrisNet and JWebDust provide applications or framework components with a database view of all sensor network systems. Most others follow a centralized broker structure in which a central entity takes care of interactions with different SAN systems.

Table 5.2 provides a summary of the functional features of the surveyed approaches. Most notably, nearly all proposals fail to address **accountability and access control** of service interactions between application and SANs as well as **privacy, trust, and reputability** of offered SAN services and information. Only Urban-sensing considers access control, privacy, and data integrity as fundamental issue that needs to be addressed as part of the architecture, but current results are limited to mere conceptual discussions. Nearly all approaches provide some way of **service discovery** to the applications. **Service composition**, if provided is preliminary static, mostly based on information available to application at service discovery time. Most approaches do not address issues of **mobility and sudden service unavailability** with respect to longer-lasting queries to ensure the continuity of request information and actuation services. Similarly, closely related **mechanisms for ensuring quality of information (QoI) and actuation (QoA)** are not provided by nearly all solutions. Only SWE and SenseWeb enable QoI attributes to be attached as metadata to sensor readings or information which can be used as selection criteria during service discovery. Few proposals allow **intermediate in-flow processing** services to be accommodated between the SANs and applications. This is essential to facilitate high-level composition of context information and semantic adaptation to be performed, which can even involve information coming from different sensor networks. The surveyed approaches provide limited or no support for **resource arbitration**. In particular for actuation services, mechanisms to manage concurrent access of resources are essential, but all proposals fail to explicitly address these requirements with specific solutions. Finally, many of the surveyed approaches fall short in addressing **management support** within the framework. Only CoBis, JWebDust, and USN provide some management tools within their framework, which address some aspects of the overall system.

5.2.2 Prevailing Sensor Networks Integration Frameworks

SNIFs can be classified according to the system architecture into two main groups: server–client SNIFs and peer-to-peer SNIFs. The former type of integration frameworks can be described as a central system which requires data owners to register their data sources to one central server. These sensing resources are updated periodically to let the server know about their availability. When an application submits a query to search for a service, the central server analyzes the query and finds the appropriate sensor network, and then produces a response. The latter class of SNIFs adopts P2P techniques where each WSN with a gateway acts as a peer. The main goal of P2P overlay is to treat the underlying heterogeneous WSNs as a single unified network in which users can send queries without considering the details of the network. User peers communicate with gateway peers in a P2P approach.

Table 5.1 Summary of nonfunctional properties of the different WSN intergration approaches

	IrisNet	Hourglass	Janus	JWebDust	SWE
Basic concept	Database view	SANs as services in a stream processing overlay	Centralized broker	SQL DB backed with framework services	Web-service-based standards framework
Framework structure	Distributed, two-tiered	In principle distributed but not realized	Centralized broker	N-tier (5)	SPS acts as central-ized broker
Data representation	User-defined DB schema per application	Topic and predicate per service	Exported as functions at SAN	Basic sensor types	XML-based descriptions
Heterogen. of SANs	Yes	Yes, support for low capability by proxies	Yes	Yes, limited to TinyOS support	Yes, SOS facilitates unified interface
Flexibility of queries	Limited to database schema/ XPATH selections	Limited to HCDL expressiveness	Limited to set of functions provided at a SAN	Limited to SQL-like queries over sensor types	SOS allows simplistic queries
Scalability	Medium	Medium, large signalling overhead for circuit management	Low, due to centralized nature	Low-Medium, all data have to go through centralized DB system	Medium-high, direct WSN access; however, centralized broker
Interaction types	Query, streaming, event-based	Streaming	RPC-like interactions, query and event-based	Query, streaming, event-based	Query, streaming via SOS, event-based via SAS/ WNS
Level of Mediation	Low, exact knowledge required (schema)	Medium, service discovery and CM	Medium, provides discovery of APIs	Low, may implement broker in middle tier	Medium, SPS
Implementation available	Yes	Yes, basic functions, single domain	Yes	Yes	Yes, components

(continued)

Table 5.1 (continued)

SenseWeb	GSN	e-SENSE	Urban	USN	Cobis
Centralized broker, web services	Sensors as service, container infrastructure or virtual sensors, stream processing	Service enabler and GW, IMS-based	High level framework	Service enabler and GW, IMS-based	SoA-based enterprise system integration, Distributed middleware
Centralized broker	Distributed peer	Centralized broker, pub-sub	Unkown	Centralized broker	SoA middleware
Unknown	XML-based description, key-value predicates	XML based description	Unkown	XML-based (SensorML)	XML-based (CoBIL)
Yes, unified webs service interface	Yes, by the help of wrappers	Yes	Yes	Yes	Yes
Medium-high, provides metadata-based selection	Will depend on how services are located	XML-based queries, depending on knowledge base in SE	Unkown	XML-based queries	XML-based queries
Medium, data flows through centralized broker	Medium-high, depending on support infrastructure	Medium, data have to go through SE	Unkown	Medium, centralized SE in data path	High as completely distributed
Query, streaming, event-based	Query, streaming, event-based	Query, streaming, event-based	Unkown	Query, event-based, streaming	Query, event-based, streaming
Medium	Low	Medium-high, broker provides matching	Limited, Mediation in terms of network enforcement points	Medium-high	Depending on the tools
Yes	Medium	Yes, lacks semantic query process	No, currently only concept	Yes	Yes

Table 5.2 Functional properties of the different WSN integration approaches

	IrisNet	Hourglass	Janus	JWebDust	SWE
Access control	Based on senselet	No	No	Unkown	Unkown
Accountability	No	No	No	No	No
Actuation	No	No	No	No	No
Caching & history	Yes	Yes via operators	No	Yes	Yes, data repositories
Fault tolerance	Replication of distributed DB	Disconnection	No	Disconnection	No
Geo-support	Yes	Yes, via predicates	No	Unkown	Yes, geo-tagging
Mobility support	No	No	No	No	No
In-flow processing	Only at sensor source	Yes	No	Possible in middle/ presentation tier	No
Privacy	Privacy filters on senslets	No	No	No	No
QoI, QoA	No	No	No	No	Limited, allows QoI description of sensor information
Traffic Optimization	Within the overlay of framework	Yes, between overlay nodes	No	No	No
Resource arbitration	Yes on SAN level	No	No	Possible in middle tier; however, not implemented	No
Service composition	Static	Semi-static, only SEPs may be replaced	No	Possible in middle tier	Limited and centralized, SPS
Service discovery	No	Yes	Yes	Yes	Yes
Trust and reputability	No	No	No	No	No
Management support	No	No	No	Yes, web-based tools	No

(continued)

Table 5.2 (continued)

SenseWeb	GSN	e-SENSE	Urban	USN	Cobis
Policies at GW	Yes	Yes, but simplistic	Mediator as policy proxy	IMS-based	No
No	No	No	Unkown	No	No
No	No	No	No	Yes	Unknown
Yes	By use of inter-mediate services	No	No	No	No
Disconnection through mobile proxy	No	No	No	No	Unknown
Yes, geo-tagging	Yes, geo-tagging	Yes	Space–time coordinates	Yes, geo-tagging	Unknown
Mobile proxies for SANs	No	IMS-based for SAN	Limited, change of mediator supported	IMS-based for SAN	Unknown
Data transformers btw broker and application	Yes, limited to SQL-like manipulation of I/O streams sync of data streams	Within SAN or SE	Possible, but not defined	Event filtering and processing in SE	Unknown
Limited through metadata	Encryption between containers	Simple A&A	Mediator as privacy proxy	No	No
Limited, allows expression of QoI metadata	No	No	Quality checks of data?	No	No
Yes, data combination across queries in broker, caching	No	No	No	No	No
No	No	No	Via mediator	No	Unknown
Static	Static	Dynamic in SE, not implemented	No	No	Yes, own language
Yes	No, provides metadata for discovery of virtual sensors	Yes	Yes via registries	Resource discovery	No
No	Data integrity	No	Yes	No	No
No	Life-cycle and resource management of virtual sensor in containers, time service	No	No	Yes via SE	Yes, support tools

5.2.2.1 Server–Client

SenseWeb

SenseWeb [4] provides an infrastructure for sharing information generated by globally distributed sensor networks. Applications can use SenseWeb to create a variety of different applications, otherwise not possible due to the lack of sensor network coverage and diversity of sensor information. Heart of the SenseWeb system architecture is the *coordinator*, which is central point of access into the system by all applications and sensor network contributors. It can be seen as a centralized broker that coordinates the information access of application to relevant sensor networks. The coordinator is decomposed into a *tasking module* and a *senseDB*. The tasking module receives application requests and tries to find matching sensor network information, considering required accuracy, capabilities, and policies of available sensor networks. The senseDB component of the coordinator tries to optimize data access across different application queries with overlapping space–time window by combining requests for common data whenever possible or serving request from cached data of previous queries. The senseDB also indexes sensor network characteristics and other shared resources in the system and enables their discovery by applications. Sensor or sensor networks are connected via *sensor gateways*, which on one side implement sensor-network-specific access methods, but on the other side, expose a standardized WS API to allow other SenseWeb components to access sensor data streams, submit data collection demands, or access sensor characteristics. Sensor gateways typically implement policies defining what sensor information is to be shared. Sensors that do not have a gateway can be connected by a shared gateway referred to as *Datahub*. In addition, *mobile proxies* are special GWs dedicated to one spatial area that allow mobile sensors to opportunistically provide information, while hiding the temporary availability of different sources to applications. Senseweb also provides data transformers that convert data semantics by some processing. *Data transformers* can be shared across multiple applications and link themselves between applications and coordinator.

SenseWeb has a variety of features that are able to deal with heterogeneity and scalability present in the real-world Internet. Heterogeneity in sensor network access is overcome by providing access through a unified WS interface. Heterogeneity in terms of sensor information quality and access policies is addressed by metadata in the sensor descriptions and learning sensor characteristics (e.g., disruptions of availability) at runtime, while allowing application to explicitly specify their requirements. Improved scalability is achieved by minimizing data collection for common data among different application queries and approximating subset of information, e.g., based on cached information. In addition, SenseWeb allows data to be collected only when actually required by applications. While these features certainly contribute to scalability by reducing the amount of generated traffic in the system, the architecture of SenseWeb does not scale well for many applications and many sensor networks due to its centralized broker nature. While SenseWeb provides support for inserting data transformers, between the coordinator and applications, it does not seem to provide means to insert such mechanisms closer to the source between coordinator and sensor networks, where they often make more sense. Continuous queries are static in a sense that once the coordinator decides on a way to serve a sensor request, they cannot adapt to changing availability of services. Only mobile proxies provide a limited support to deal with such changes at sensor network level. While mentioning incentives, cost-sharing, security, privacy and trust, SenseWeb does not provide explicit support for such functions in its infrastructure.

Janus

Janus [6] is an attempt to break up the tight coupling between sensor networks and their application, by inserting an intermediate broker into the interaction path. Janus makes use of extensible resolution protocol (XRP) and introduces two entities as part of its architecture, namely an *XRP agent* and an *XRP engine*. Instead of directly interacting with a sensor network, applications interact with an intermediate broker realized by the XRP agent, typically located somewhere in the access network. The XRP agent then interacts with *XRP engine*(*s*) located in the gateway of to the sensor network(s), via an RPC-style interface. The XRP agent can discover available services at a sensor network and gain access to the services by receiving a locator bound to local functions calls at the corresponding XRP engine. These locators can be used as *selectors* to identify services in subsequent RPC function calls at the remote sensor network, realizing both query–reply-based as well as event-based interactions. Different applications can be interfaced to the XRP agents via the implementation of application specific proxies.

Janus achieves a decoupling of sensor networks and applications, by introducing itself as level of indirection between the two systems. Thus, applications and sensor networks can evolve independently, while relying on the Janus framework to remain an invariant achieving compatibility via XRP. Janus is able to integrate different heterogeneous networks, as long as they implement an XRP engine that exports the available services via new RPC selectors and that implements a translation of the function RPC calls to the sensor network native mechanisms. Heterogeneous applications can be supported; however, for each application a specific proxy needs to be implemented that interacts with the XRP agent. Clear interfaces to the XRP agent seem not to be specified, which makes it difficult to write application proxies. The reliance on an XRP agent as centralized broker makes Janus not scale well for large number of applications and sensor networks. Janus does not provide any support for composition of context information of different sensor networks or functions to aid the automatic selection of appropriate sensor networks for interaction. It shifts the onus to perform this task to the application. Janus does not provide any mechanisms that are able to optimize the delivery of same sensor information to multiple applications, nor does it address functions for security, privacy and trust, and accounting.

JWebDust

jWebDust [15] provides a software framework that allows web-based applications to query and control multiple potentially heterogeneous wireless sensor networks. WebDust is based on a multitier architecture, splitting the overall system into five tiers, namely sensor, control, data, middle, and presentation tier. All framework components are implemented in Java, apart from the ones contained in the sensor tier. The *sensor tier* is formed of one or more wireless sensor networks consisting of sensor nodes (motes) operating a TinyOS-based jWebDust firmware. The jWebDust firmware enables multi-hop routing within the WSN and provides support functions such as query subsystem, discovery services, monitoring service, and time synchronization. Sensor networks are connected to control centers that form the *control tier*, acting as gateways between the sensor tier and data tier. The control centers are responsible for the gathering of all readings coming from the sensor network and the forwarding of queries from the data tier to the sensor nodes. Control centers periodically poll the data tier for new available queries and store all available sensor readings into the data tier. Control centers are able to handle temporal disconnections of the sensor tier from the data tier by buffering of sensor readings until reconnection subject to local capacity constraints. The *data tier* is based on a relational database system and hosted by SQL servers. Information is stored in tables that can be grouped into three categories. Mote-related tables store information on the hardware characteristics and sensing services of sensor networks. Query-related tables hold information on currently active queries in the jwebDust system. Sensor-reading tables hold sensor readings that have been performed by each particular mote in the system. The *middle tier* provides a set of reusable components that allows the mapping of the information stored in the tables of the data tier, the manipulation of the information, e.g., creation of queries in the data tier and implementation of rule-based actions and notifications. The *presentation tier* implements the user interface components visible to the end user, providing interaction controls and sensor data visualization tools.

jWebDust decouples applications from the sensor networks via the *n*-tier architecture and supports horizontal composition of WSAN applications of different domains. Applications can learn about the availability of different sensor networks and their capabilities by querying/browsing the data tier. Individual motes within a WSN are assumed to have unique IDs. In order to make them globally distinguishable across multiple WSNs, each WSN is assigned a unique sensorNetworkID. The addition of a new sensor network requires some initial configuration, e.g., assignment of sensorNetworkID and inclusion of novel mote and sensor types. Afterwards, discovery of new motes and sensors in a sensor network executes automatically. jWebDust supports interactions with different heterogeneous WSNs, where control centers are able to hide the heterogeneity of WSNs from the data tier. Although the framework claims to support a variety of different query types, applications currently require explicit knowledge of a sensor network and sensor types (learned from the data tier) in order to create various queries for sensor information. Service broker-like components with semantic query support could theoretically be implemented as part of the middle tier; however, such functionality is currently nonexistent. Although jWebDust claims to provide concurrent access by multiple applications, mechanisms for resource arbitration are currently lacking. JWebDust also requires all sensor nodes to implement the same firmware for correct query processing and service discovery. While most TinyOS-capable platforms are supported, it requires each sensor node in the network to be reprogrammed and configured prior participation in the framework. jWebDust also lacks functions for security, privacy and trust, and accounting and provides no explicit support for controlling the access to different sensor network resources. The reliance on a central relational database in the data tier can become a scalability bottleneck, once the number of participating sensor networks and querying applications grows.

IrisNet

IrisNet [4] is one of the first attempts to develop an architecture that is able to provide integrated access to globally distributed sensor networks over the Internet. The IrisNet's goal is to reuse the infrastructure of deployed sensor networks by enabling the sharing of generated sensor feeds among many applications (sensing services). IrisNet provides sensing services with the view of a distributed database in which data of different sensor networks can be collected and queried. IrisNet is realized as a two-tiered architecture with *organization agents* (*OA*) and *sensing agents* (*SA*) as fundamental components. A developer of a sensing service provides a database schema tailored for its application, which is implemented on a possibly distributed set of OAs. The group of OAs maps to a single sensing service and must collect and organize sensor data to answer a particular set of service-specific queries. OAs form a distributed database in which data are hierarchically organized in self-describing tags based on XML (to naturally reflect the hierarchal organization of existing geographic and political boundaries). Queries on the database are expressed in XPATH and select data from a node set in the hierarchy. Each of the distributed OAs can store a subset (subtree) of the hierarchy in which each node either points to data sources for sensor streams (represented by SAs) directly providing corresponding service data or other OAs, besides pointing to other nodes, that implement parts of a missing subtree. The IrisNet infrastructure enables the distribution of the query to adequate OAs and ultimately to the SAs providing the required data, and the composition of the final response across multiple OAs on the reverse path. OAs register a global name and IP address with DNS, so queries across distributed OAs can be dynamically resolved. Data of queries can be cached at their corresponding OAs, and repeated requests directly served from cache to improve subject to freshness requirements of the sensing service. IrisNet also provides replications of OAs and placement of OAs as additional mechanism to improve system reliability and query performance. SAs provide a generic data acquisition interface toward sensors and sensor networks and typically collect raw sensor readings as required. Besides a database schema for the OAs, developers of sensing services write so-called senselets that execute in a secure environment of the SAs. These senselets are able to process (e.g., filter) the incoming raw sensor stream and send the processed sensor information to nearby OAs. In addition, SAs mediate the access of senselets to the resources of its attached sensors.

The IrisNet architecture seems to provide several desired features, ranging from sensor network reuse, application-specific in-network processing resource mediation on the SAs, fault tolerance, and geographic information lookup and seems to scale well as it provides a distributed database view for each sensing service. While providing the possibility of sharing computation across senselets, it does little in optimizing the data traffic from the sensor networks to the sensing services. Data is routed via the OA overlay, which may result in suboptimal data paths (could be alleviated by OA placement if physical hosts are arbitrarily available) and it does not allow concurrent applications to share the same sensor data. Creating a distributed database for each sensing service may lead to services often implementing redundant databases that could have been shared among several applications. IrisNet also does not provide discovery mechanisms that allow sensor networks and their capabilities to be automatically discovered by application developers at design time, not to mention runtime. IrisNet also does not address functions for security, privacy and trust, and accounting.

Ubiquitous Sensor Networks

The work presented in [16] represents the first step of an ongoing research activity Telefonica is performing toward the Ubiquitous Sensor Networks concept from the ITU-T [23]. The presented platform is being designed following a horizontally layered approach, so networks and services can evolve independently. The four layers of the platform, following a bottom-up approach, are: the Sensors and actuators networks, the Gateway (that provides independence from the networking technology), the NGN core (IMS), and the Service Layer (where an enabler is provided). The key elements of the platform are:

- The *USN-Gateway*: is a logical entity whose main goal is to provide independence from the sensing or networking technologies used to communicate sensors and actuators. The independence is provided by performing two transformations: from one side it provides homogeneous communication toward and from the sensors and actuators networks, and from the other side it provides homogeneous data representation. It is being defined as an IMS User-equipment which already provides important functionalities like AAA and it allows to be deployed in a wide range of devices.
- The *USN-Enabler*: is defined as an OMA enabler, intended to allow services to be created in a cost-efficient way following a horizontal approach, where multiple services can access the same sensor and actuator networks. The basic functionalities it provides are: resource discovery, publish–subscribe–notify mechanisms, event-filtering and processing, and homogeneous remote management.

More than the functionalities it provides the key issue of the USN-Enabler is the way in which it has been designed, since it follows the OGC Sensor Web Enablement Family of Standards and the OMA Presence Simple and XDM specifications.

- *Standardized homogeneous representation of sensor data and metadata*: It provides homogeneous representation of the sensors and actuators representations and measurements following the OGC® SensorML and Observations & Measurements (O&M) standards.

The more interesting issues brought by the platform, more than the functionalities it provides, are the way in which these functionalities are provided, since this approach, instead of redefining some existing functions, uses the existing standards. Especially important is the use of SensorML as the language that unifies the heterogeneous sensors and actuator definition. Even considering that the USN Platform presents an interesting approach to tackle the problem of integrating sensors and actuators to services, mainly due to the extensive use of standards, it is still a first step and much work still needs to be carried out for it to be considered as a solution for the USN. Issues like billing, trust, accounting, and high-level interaction mechanisms are not still attached. The architecture can be viewed as centralized, but with catalog functionality in order to have some distribution of functionality. It is mentioned that some of the functionalities could be provided by elements like the Gateway, but it is not yet defined.

e-Sense

The framework presented by the e-SENSE approach [17] aims at integrating sensor networks into the IP multimedia subsystem (IMS) of future mobile and converged networks. The framework allows applications in IMS-based service platforms to access sensor and context information from a variety of sensor networks with heterogeneous capabilities. The framework introduces two architectural components, a context service enabler and gateway extensions to sensor networks. The context service enabler provides sensor-based context information as a dedicated service via a unified interface using standardized IMS protocols (such as SIP). Thus, the context service enabler can be used as a service building block for the realization of various different context-aware applications hosted on application servers in the IMS domain. Gateway extensions allow the integration of heterogeneous sensor networks into the IMS domain by implementing sensor-network-specific mechanisms on one side and provide service functions for interaction with the context service enabler on the other. Each sensor network gateway is an IMS user identified by a unique IP multimedia private identity, several public ones for each gateway. Using the proposed gateway extensions, the sensor network systems register their presence with the IMS core platform and available services with the context service enabler. It is expected that applications and services will express their context requirements on a semantically high level of abstraction and that the context service enabler may have to interact with multiple sensor network systems to derive the required context information components. Based on an incoming context information request of an IMS application context service enabler performs a decomposition of high-level context information requests into low-level service task graphs to be executed by one or more wireless sensor networks. It then requests the low-level services from the identified sensor network gateway and composes the context information response from the required context information components of the service responses. Communication between sensor networks and gateways is enabled by publish/subscribe mechanisms based on the SIP event framework.

The e-SENSE framework offers several desired features such as discovery of available sensor networks and their service capabilities. It enables a decoupling of application from the underlying sensor networks – that is, applications can enquire the contextual information without requiring knowledge of the underlying sensor networks. Application queries can encompass information that may be jointly provided by multiple sensor networks and processing and composition can be achieved either in the sensor networks/gateways or in the service enabler. While making use of scalable well-understood signaling mechanisms, it raises scalability concerns as communication requires always the involvement of the centralized service enabler. Although multiple physical instances of a context service enabler can be deployed, the architecture does not address how coordination between different such instances is achieved. Optimization of data flows across different queries are currently not addressed by the proposed frameworks, as are not adoptions of existing queries to changing conditions in the system. Furthermore, the architecture of the e-SENSE framework falls short in considering security, privacy, trust, and accounting issues.

5.2.2.2 Peer-to-Peer

Global Sensor Networks

Global Sensor Networks [18] is an approach of providing a distributed middleware platform for integrating heterogeneous sensors into a "sensor web" providing internal stream processing capabilities on the exchanged sensor information. The architecture of the GSN framework is based on distributed peer entities called GSN containers. *GSN containers* are typically deployed at normal Internet hosts or servers and communicate with each other via point-to-point connections. Core element in GSN is the so-called *virtual sensor* abstraction. Virtual sensors abstract implementation details to access information from physical sensors and allow a unified way of treating sensor services or composed sensor service by the middleware. The specification of a virtual sensor includes metadata for identification and discovery, the structure of input and output streams, SQL-like internal streaming processing and properties related to life-cycle management and physical deployment. Virtual sensors have one or multiple input streams and produce exactly one output stream. Input streams can come from physical sensors interfaced via implementation-specific wrappers or other virtual sensors. Virtual sensors can manipulate and combine streams of different characteristics using SQL-like operations and conditionally produce output streams (event-like). Virtual sensors can be dynamically deployed on the GSN containers and the production of its output stream is dynamically triggered by the arrival of input streams. Besides running instances of deployed virtual sensors, GSN containers provide additional functionality supporting the management of the virtual sensor instances and their required resources, function to manage streams and resources required for stream processing, query management (request input data from other virtual sensors and keep track of other virtual sensors requiring their output), and a storage layer for the management of persistent storage of data streams. Access to GSN-container internal functions is provided by an interface layer, which is used to communicate between GSN containers or can be accessed directly via web interfaces. Besides providing access control at different levels of granularity down to the virtual sensor level, the interface layer provides integrity and confidentially functions for the exchange of data streams.

GSN provides many interesting features. The ability to create aggregate virtual sensors from different heterogeneous sensor information sources suits well the context information processing demands. Unlike other surveyed frameworks GSN offers access control to sensor information and integrity protection. In addition it provides a plug-and-play-like feature for integration of new sensors which allows upon detection of a new sensor the dynamic download of a IEEE1451 transducer electronic data sheet and automatic generation of a virtual sensor (given the wrapper code for the WSN technology is known). A current limitation represents the SQL-like stream processing operations, which do not allow complex processing data stream and data fusion to be performed in the system. The decentralized peer-to-peer nature of the system seems to indicate good scalability properties. It is unclear however, how data streams between GSN containers can be optimized, e.g., by selecting the same virtual sensor streams for different independent queries. Point-to-point transmission of streams between GSN containers may be another feature potentially reducing its scalability. Although meta-information is provided for virtual sensors, which can be used for service discovery, it is still unclear what mechanisms would be used to discover virtual sensors. The current descriptions seem to "hard code" required input streams into the virtual sensor descriptions, which does not make dynamic composition or late binding of virtual sensors possible at runtime. GSN does not seem to provide any infrastructure support for accounting nor does it provide a little information on an infrastructure that can be used for dynamically composing or modifying active virtual sensor services.

Sensor Web Enablement

Sensor Web Enablement (SWE) [19] is an initiative by the Open Geospatial Consortium aiming at the development of a set of standards to enable the discovery, exchange, and processing of sensor information and tasking of sensor systems over existing Internet. SWE strives for plug-and-play-like integration of sensor networks and enabling protocols to make those accessible and controllable by web-based applications. The current standards framework encompasses seven different standards, some of them completed and others in draft stage. Three of the standards are concerned with the XML-based encoding and representation of sensor information/observations and the description of sensor capabilities and related information processing steps. The remaining four standards describe standard web service interfaces for tasking of and interaction with sensors. The *observations and measurement* (*O&M*) schemas provide XML schemas for representing and observations, measurements, procedures, and metadata of sensor systems and efficiently encoding them for transfer and archiving. The *sensor model language* (*SensorML*) supports the description of a functional model of a sensor system by providing models and XML schema for describing processes of measurement and post-measurement processing and their exact chaining. The *transducer markup language* (*TML*) provides models and CNK schema for describing hardware response characteristics of transducers (more complex integrated sensors/actuators) and efficient method for encoding and real-time transport of sensor data. While partially overlapping with SensorML, TML focuses more on support of streamed real-time sensor information flows, preserving their spatial and temporal association for later data fusion. The *sensor observation service* (*SOS*) specifies a web service interface that allows SWE clients to obtain observation and measurements from a collection of sensors. The SOS also allows clients to access metadata information about associated sensors, platforms, procedures, and other metadata associated with observations. The information is exchanged using the three aforementioned XML-based data formats. The *sensor planning service* (*SPS*) acts as a broker service between clients and different SOS. It allows clients to determine the availability of certain sensing services that may be needed to satisfy collection requests and the feasibility of those via a standardized web service interface, potentially spanning multiple sensor systems and management of such collection requests. The *sensor alert service* (*SAS*) provides web service interfaces that allow clients to subscribe to alerts/event notifications of particular sensors. The SAS acts only as a registry that enables clients to determine the nature of available alerts, protocols used, and the options to subscribe to specific alerts. Alerts or event notifications themselves are forwarded my messaging servers. The web notification service (WNS) specifies a web service interface that allows clients to interact with one or more services in an asynchronous way. WNS provides support for both unidirectional and bidirectional asynchronous communication.

The standards framework of SWE addresses many issues including standardized descriptions for sensor/actuation platforms, actual sensor information and processing chains in a sensor web as well as several interfaces for applications to interact with sensor systems, perform asynchronous communication, and manage event notifications. In addition the SWE framework defines interfaces for a service-broker-like component, the SPS, which enables more complex interactions between applications and sensor systems, such as determining suitable sensor observation components across multiple sensor systems and information repositories to satisfy more complex sensing requests and the respective tasking of the sensor systems. While specifying the high-level architectural framework and interfaces, the SWE does not address the realization of the framework services or required interaction protocols.

Some of the standards are at draft stage and still undergoing specification and further harmonization between the different standards brought into the framework from outside, such as SensorML and TML is required. The framework does not explicitly address aspects of security, privacy, trust, accounting, and resource arbitration. In addition, the SWE representation formats allow application only to express simplistic queries and are not suitable for the high-level declarative service interface as well as the complex processing of sensor information inside of the system based on ontological models.

Hourglass

Hourglass [13] aims at creating an Internet-based framework for connecting heterogeneous geographically distributed sensor networks with applications that require sensor information. Hourglass provides an infrastructure for data collection referred to as data collection network, which handles service naming, service discovery, route setup from sensor networks to applications, and provides support for integrating internal services along the data dissemination path to perform aggregation or buffering of sensor information. Hourglass primarily addresses stream-based aggregation and processing of sensor information that is required by applications over a longer period of time. Hourglass treats the sensing and processing capabilities sensor networks offer as services, and extends the service concept to also encompass any intermediate processing service on sensor data. Typically services can act as data consumers, data producers, or both. Services in Hourglass are organized into service providers. Each service provider comprises more Hourglass nodes forming a single administrative domain, entering or leaving the Hourglass system as a unit. Each service provider needs to support minimum functionality in terms of a *circuit manager* and *a registry*. In addition a service provider can provide several generic or application-specific services. A service registry is a (distributed) repository of information about various services and active circuits in the Hourglass system. It is a lookup service that allows the resolution of service endpoints. Each service provider typically maintains an own local registry, with which active services of a service provider register via service announcements. Such service announcements typically contain communication endpoint identifiers, topic name, predicates, and expiration time as entries are kept as soft state. An application that aims to establish a "streaming session" with one or more sensor information sources and intermediate processing first queries the service registry for available services in the Hourglass system. It then specifies its query requirements as so-called circuit descriptions that link one or more data producers and a data consumer with possible intermediary in-network services into a logical data flow. The circuit manager instantiates the described logical flow as network data flow by establishing connections between the different physical nodes offering the respective services. The hourglass service layer manages the invocations to the service interface and the multiplexing of data to and from the connected circuits. Sensor data are routed along the established path and possible processed at intermediate nodes.

Hourglass provides reliability to system dynamics by explicitly supporting a mechanism to deal with temporary disconnections of a circuit, that is, if the connectivity to a service provider that is part of a circuit becomes unavailable. Disconnections are monitored by heart beat mechanism along the circuit, based on explicit control messages or implicitly by data that are exchanged. Once a disconnection is detected appropriate actions such as buffering of data can take place in the circuit. Thus, Hourglass offers the advantage to modify existing circuits to adapt the services to changing conditions for continuous application queries. It allows optimization of the delivery of same sensor information by combining transmission between service endpoints across multiple circuits/applications. While the architecture has the intention to scale well by planning for distributed operation across multiple service providers, it leaves open how respective service registries distributed across multiple service providers interact or are managed and how the connection managers of different service providers cooperate to establish connections across multiple domains. The overhead for establishment of circuits for each data request by application together with the fact that state needs to be maintained at each node that is part of the circuit is a severe scalability concern. While it may be justified for streaming-type queries for longer periods of time, it does not suit well one-shot queries or periodic queries with little data exchange. However, the framework does not address functions for security, privacy and trust, and accounting.

Urban Sensing

In the Urban Sensing project [5] they consider three types of applications: personal, social, and urban. A personal application uses information about the end user for the purpose of the end user. A social application mimics Facebook and other social networking sites, where data are shared among a set of users for free. In urban applications the users share data with the general public, and the importance of identity control, etc., is thus much higher. It is argued that new network architecture is required in order to share data in a controlled way and to assure basic quality checks of data. In this the authors see an evolution from single-domain WSNs to collective/federated WSNs to full integration into the full global infrastructure. The federation of WSNs is referred to as the sensor fabric.

In order to achieve full integration, the global network must know about the abstractions used in forming the sensor fabric, and the sensor fabric must import notions about the future global network into itself. It is hence a two-way process.

Abstractions required to form the sensor fabric could be of the following types:

- Space–time coordinates
- Policy-mediated rendezvous based on data properties and metadata
- Aggregation-based reliability

The authors argue that embedding these abstractions into the global network changes the network from *host-centric* to *data-centric* in nature.

The authors subsequently argue that most important to solve while incorporating sensors into the global network are the issues of verification, privacy, and dissemination.

The authors therefore seek to incorporate embedded basic data protections into the fundamental mechanisms of the network. The proposed architecture incorporates the four entities below, and also existing network services such as trusted Certificate Authority (CA).

- *Sensors*: These are data sources at the edge of the network; these are not simply a pure source of data, but can also provide a control point to the external world. These control points could be for the purpose of configuring the sensor or for providing global contextual information.
- *Subscribers*: These are the users of the data provided by the sensors; individual users of data or applications providing some value-added service.
- *Registries*: These are network services that help subscribers to find and bind with sensor data streams. Sensors register here and subscribers query these in order to find the sensors they want. The type of handle provided by the registry is extremely important.
- *Mediators*: Nodes in the network that provide selected in-network functions on sensor data streams. These services could be to perform verification of data streams or to provide anonymization of sensors to subscribers.

In a typical deployment scenario, a sensor owner registers a sensor in Registry 1 via the mediator Mediator 1. Registration contains sensor type, location, and context + disclosure and verification rules. If sensor is mobile, it may change mediator over time. The sensor then initiates data transmission to Mediator 1 – either on demand or proactively depending on configuration. The role of Mediator 1 would be to act as a privacy proxy and to provide a network testimony of the validity of the context of the sensor.

A subscriber then sends a query to Registry 1 via a mediator Mediator 2. The query has to go through a mediator since the sensor may have privacy rules depending on the context of the subscriber, and Mediator 2 attests this context. The registry then returns a pointer to the data streams. Mediator 2 can then bind to these data streams, which means that in this case it binds to Mediator 1. Mediator 1 can now run its own privacy rules and allow/disallow sensor data access.

The urban sensing architecture takes a very protective stance on participants, and argues that network support is required for verification and dissemination of data. By embedding support for these functions into the network, it is easier to efficiently and securely execute them. Many issues around management of sensors/sensor networks are, however, not touched upon, and the underlying efficiency of processing/context sessions are hinted at. The basic security architecture is only sketched.

CoBIs

The CoBIs project [20] developed a radically new approach to business processes involving physical entities such as goods and tools in enterprise environments. Advances in networked embedded systems were applied to embed business logic in physical entities to create so-called Collaborative Business Items (CoBIs). Such items enable to relate more closely the state of an enterprise as represented in a business process with what is actually happening in the real world. Thereby, business processes can be extended to the "point of action" rather than via a centralized back-end system.

The central concept of the CoBIs project was to use a common service paradigm throughout all layers, from the enterprise application down to the logic executed on sensor nodes. A middleware was built based on a service-oriented architecture (SOA). The middleware allows the deployment of business logic in the form of services to the edge of the network and onto the sensor nodes themselves. CoBIs were focused on providing the basic SOA framework as well as the tools to monitor and manage the network. Using a service-oriented architecture in the context of distributed embedded devices as well as sensor and actuator networks solves several problems that are usually associated with such systems. Solutions concern especially, the integration of sensors and actuators with enterprise systems as well as the management, monitoring, and administration of a system with highly distributed logic.

In addition to the SOA framework, a set of reusable collaborative services was defined and described in a newly developed service description language called CoBIL. A CoBIL service description includes a definition of the interface, which is based on the Web Service Definition Language (WSDL). Furthermore, it includes a textual description of the service as well as information about the composition of the service and technical constraints for the deployment.

Three different sensor network platforms, namely Particles, μNodes, and Sindrion, were integrated with the middleware through a common abstraction layer to demonstrate the feasibility of connecting heterogeneous hardware to the system [24]. The different platforms have different characteristics: Depending on the application scenario, one can thus choose the most suitable technology. Criteria have been developed that will help end users to make that choice, also comparing it to existing technology like RFID and wired sensors.

While the middleware, service description language, and system support tools developed could be the foundation of a widespread, multi-partner sensor network infrastructure, CoBIs did not address formal semantics and context models or security issues.

5.3 Road Ahead

5.3.1 Introduction

Internet that we know today was designed 40 years ago as a tool that will facilitate easier exchange of information between researchers. From that vision, Internet grew to a ubiquitously available platform people and businesses depend on in all aspects of everyday life: social networking, business applications, health care, learning, information exchange, etc. With proliferation of mobile networks and particularly with the introduction of high-speed mobile technologies (HSPA – High Speed Packet Access), the requirements for Internet access broadened from just home and office environments to any place and at any time, including while on the move in a car, a bus, or a train. In addition to that, it is becoming a norm to have private data stored online to make it always accessible. The pervasiveness of Internet and its intertwining with everyday life has brought a number of new requirements as well as problems together with the benefits it provides. Security and data privacy present a big problem, with spam email, scams, and identity thefts contributing a huge percent to the overall Internet activities.

In order to efficiently support these new trends, a number of activities have been initiated in the last couple of years. Their intention is to design a new generation of Internet, commonly known under the name of Future Internet. In Europe, these efforts are primarily combined under the FP7 program and it is Future Internet Assembly (www.future-internet.eu). In Japan, the driving force in this domain is the AKARI initiative (http://akari-project.nict.go.jp/eng/index2.htm), in the USA it is the Future Internet Network Design (FIND) project (http://find.isi.edu) and FIF in South Korea (http://mmlab.snu.ac.kr/fif).

One of the main points that Future Internet will bring is the integration of the physical and digital worlds, i.e., embracing the Internet of Things as one of its core components. Numerous sensors, actuators, RFIDs, machines, and in general "things" will become easily accessible to other Internet users and devices, thus forming infrastructure that pervades into all aspects of our lives. This will enable efficient interaction with the physical world, adaptation of Internet applications to the users' contexts as well as influencing and changing the environment based on the applications' settings.

The "things" will range from simple sensors measuring temperature or humidity, to complex intelligent semantic systems capable of providing answers by combining a number of inputs, simple sensors, actuators, and other network services like location, security, and charging.

Before this vision becomes a reality, a number of technical, legal, socioeconomic, and business challenges and issues have to be resolved. Discovery of information and capabilities provided by different "things" in such distributed environment, standardized description of the capabilities, scalability of solutions to support huge number of connected "things," how to trust the information provided by unknown sensors embedded somewhere in the environment, how to protect privacy of "things" providing the information, new business cases on which applications and services will be built and provided, how will people react and adapt to such new Internet, etc. are just some of the challenges ahead.

In the following two sections, two projects dealing with some of these issues are described. CommonSense is an industry-driven project, focusing on integration of sensors and actuators in the mobile networks context. FP7 SENSEI is a large integrated project under the EU FP7 program with a goal to design a framework for integration of the digital and the physical worlds.

5.3.1 CommonSense

The CommonSense system [21] was proposed to enable the vision of ubiquitous sensing where sensor networks provide the missing link between the virtual and physical world. Today we observe a tremendous increase of mobile subscribers with already existing about three billion users of handheld computers. These mobile devices are powerful communication and computing multipurpose devices that are increasingly being equipped with a number of different sensors: image, sound, light, temperature, acceleration, RFID readers, etc. The ability to interact with sensors in their vicinity via built-in short-range communication interfaces like Bluetooth, in addition to the previously mentioned characteristics, make mobile devices an excellent platform for sensing the physical environment and interacting with it. The authors of CommonSense first analyze the roles of different entities which will potentially be involved in the provisioning of WSN services in the future, and they propose the system architecture that incorporates the conclusions of their study.

The first identified entity is the WSN provider who provides the sensor network its services and because of the equipment ownership, the WSN provider defines sensor network access and utilization policies. Higher-level services are provided to the end-users service providers. They combine and process different sensor networks services and other required input like Google Maps. The third entity which provides the link between two aforementioned parties is called CommonSense provider. It acts as a broker to the service providers and helps them to find a sensor network, enforces access policies set by individual WSN providers, processes the data received from multiple sensor networks before delivering these to the requesting service provider, and provides authentication, accounting, and billing functionality. The role of the CommonSense provider is to provide a unified interface to services provided by heterogeneous sensors and actuators. The CommonSense providers will collaborate with other entities such as location providers, telematics information providers, presence providers, etc. These entities, referred to as the third-party service providers, will process collected information in a specific manner or will be adding own information to the mix, thus providing additional value to the services provided by the CommonSense providers.

The CommonSense system is based on a tiered-service-oriented architecture. The service providers interact with the CommonSense provider, who in turn is the entity directly interacting with the WSNs. The service providers treat the CommonSense provider as an entity providing services, and thus have no direct knowledge or influence over how the CommonSense provider finds the appropriate data to respond to their requests. This constitutes the first level of the service architecture. The CommonSense provider then in turn treats the individual WSNs as entities offering services. This means that the sensor networks have to be able to describe themselves, where they are, and what they can offer.

Traditionally SOAs focus mainly on peer-to-peer workflow-driven processes. In CommonSense architecture the authors instead envision that individual, moving WSNs offer very thin atomic and dynamic services while the CommonSense provider offers more complex services by combining these primitive WSN services to create for example mash-ups.

The authors provide several reasons supporting this tiered architecture. The first is that they wish to create a scalable system where the focus is not on every single individual sensor, but rather collections of them offering a service. Secondly, focusing on services only the collection of sensor samples becomes independent of specific sensor network implementations as long as these networks are able to describe how the service they provide can be used. Thirdly, the authors recognize the problem for an autonomous device to manage multiple dynamic security associations and the associated authorization decisions and therefore they propose to outsource the authorization task to the CommonSense provider – something which is enabled by the tiered SOA.

The proposed architecture is mapped on three technology planes: Communication services, Application enablers, and Application plane (see Fig. 5.1). Applications are built using common service blocks residing in the Application enablers' plane and all are connected by a number of network solutions residing in the Communication services plane.

Different domains are identified on each plane. The WSN, CommonSense, and third-party service domains comprise the service plane, while the Peripheral, Access, and Core domains comprise the Communication Services plane. The application providers providing end-user applications reside on the application plane. In the Applications plane the authors differentiate between existing applications that do not depend on physical world context (e.g., call setup) and applications that cannot exist without one (e.g., burglar alarm).

The Application enablers' plane is divided into three domains: WSN domain, CommonSense domain, and third-party services domain. The WSN domain comprises all atomic sensor services, i.e., services provided by individual sensors or sensor networks and used as small building blocks of more complex services offered by the entities residing in the CommonSense domain. The CommonSense domain is where the core functionality of the proposed architecture resides. This domain does not host any specific applications, but provides a set of enablers for all types of applications. These enablers include information exchange, sensor network discovery, data processing, aggregation of atomic sensor network services (sensor mash-ups), Authentication, Authorization, and Accounting (AAA) services. In short, the domain creates the possibility of having a dynamic binding between applications and WSNs. It is mainly based on semantic technologies which provide access independency. A single attachment point for sensor networks also facilitates security

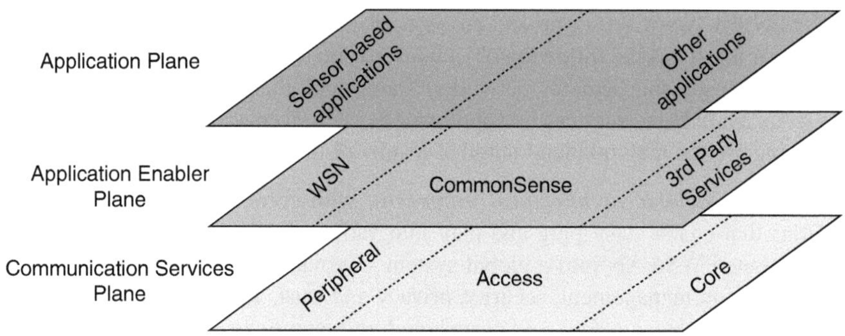

Fig. 5.1 The proposed layering in CommonSense architecture

and privacy support. The authors defined a set of needed core functionalities in the CommonSense domain. The most important function is the

Service Control Function (SCF) which controls the interaction with all external parties. High-level service requests from applications are analyzed by the SCF with the support from the Request Analyzer (RA) and sensors, sensor networks, or existing sensor mash-ups. The Request Analyzer (RA) is a decision engine that can decompose a request from an application to multiple individual information requests, and then recompose an aggregated answer. The output from the RA is used to search the SR (a database containing registration descriptions of all attached WSNs) and find WSNs with matching capabilities. Once suitable WSNs are identified, the SCF issues either standardized low-level service requests or special legacy WSN requests using a mediating function (called Service Gateway). The WSN provider functionality on the Application enablers' plane is represented by Service Gateways (SGW). The SGW represents atomic sensor network services and is responsible for mapping the SCF requests onto the sensor network technology specific commands, which is a core requirement for interoperability. The CommonSense domain entities will use third-party services as an additional tool in creation of application responses. The applications can also use the third-party services directly if they are required by the application logic. Some of the already existing services provided by the network are considered as the third-party services in the network of the future. Examples of the third-party service providers are presence servers, location providers, object identity resolution providers, etc.

The communication services plane provides the underlying secure and reliable communications services to the Application enablers and the Application plane and enables interaction of all their entities across the different domains. The Communications service plane is divided into three domains: Peripheral refers to the local connectivity functionality (e.g., Bluetooth, Zigbee), Access refers to the wireless and wired last hop connectivity functionality (e.g., WCDMA, ADSL), and Core refers to the actual backbone.

5.3.2 SENSEI

FP7 SENSEI (www.sensei-project.eu) is a large integrated project under the EU FP7 program. It has the following objectives: to create a common, global, WS&AN (wireless sensor and actuator networks) framework that will enable making the WS&AN available to services and applications via universal service interfaces. The following are the main planned tangible results of the project:

- *A highly scalable architectural framework* with corresponding protocol solutions that enable easy plug-and-play integration of a large number of globally distributed WS&AN into a global system – providing support for network and information management, security, privacy and trust, and accounting.
- *An open service interface and corresponding semantic specification* to unify the access to context information and actuation services offered by the system for services and applications.

- *Efficient WS&AN island solutions* consisting of a set of cross-optimized and energy-aware protocol stacks including an ultra low-power multimode transceiver targeting 5 nJ/bit.
- *Pan European test platform*, enabling large-scale experimental evaluation of the SENSEI results and execution of field trials – providing a tool for long-term evaluation of WS&AN integration into the Future Internet.

5.3.2.1 Initial SENSEI Architecture

The high-level overview of the SENSEI architecture is given in Fig. 5.2. The central part of the architecture is the SENSEI (Real-World) Resource layer. This layer provides a set of interfaces that applications and services use to interact with the physical world. Within the SENSEI Resource Layer, the main concept is that of a resource. A resource is a conceptual representation of any information source that enables real-world sensing or has the ability to act upon the environment and entities within it. The concept covers not just the actual entities that have direct access to the physical world, but also the entities with indirect access acquired via aggregation, fusion, or inference from other SENSEI Resources. All WS&ANs are presented by their descriptions (Resource Description) detailing the capabilities of the corresponding WS&ANs including location, access policies, available operations, type of information produced, etc.

The SENSEI resource layer interacts with the communication layer to map all requests received from the applications and services to appropriate communication channel. SENSEI community management is responsible for management of all SENSEI entities.

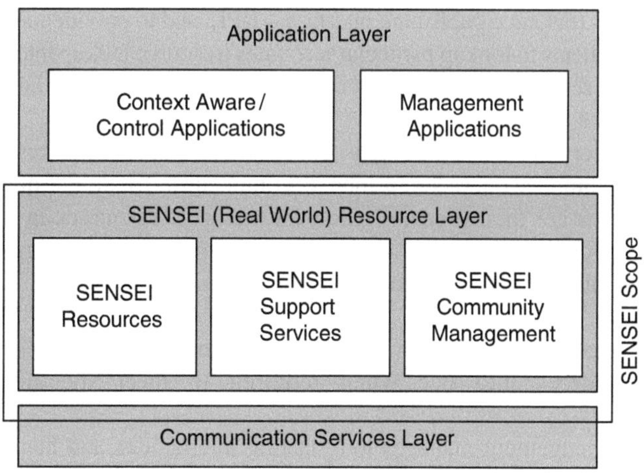

Fig. 5.2 High-level overview of SENSEI architecture

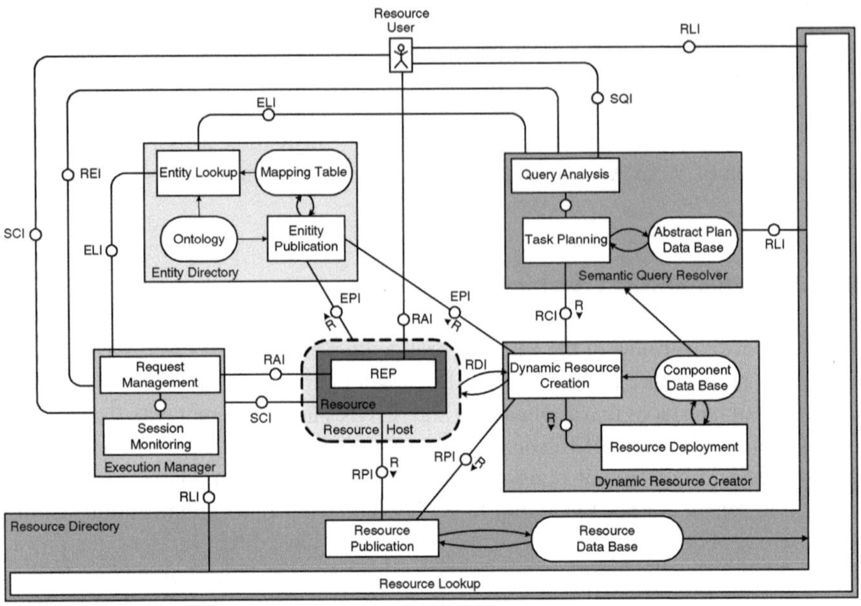

Fig. 5.3 SENSEI resource layer architecture

SENSEI Resource layer architecture is given in refrence to Fig. 5.3. The following components comprise this layer:

- (Real-world) resource provider consists of one or more real-world resources. It provides access to these resources via Resource Access interface (RAI) and is responsible for interaction with the rendezvous component to publish information about its resources.
- Rendezvous makes the glue between resource clients and resources. Its purpose is twofold: to provide mechanisms for resources to publish their capabilities and functionalities (resource publishing interface – RPI), and to provide mechanisms for the resource clients to lookup particular resources (resource lookup interface – RLI). It also stores resources descriptions of all available resources at any given time in the resource directory.
- Semantic Query Resolver (SQR) is an advanced component, responsible for analysis of complex queries and their decomposition to simple queries. These simple queries are then used to search for adequate resources in the resource directory. In case no adequate resources exist, the SQR can trigger Dynamic resource creation component to create a new resource based on the available resources.
- Dynamic resource creation is a component capable of dynamically combining several resources into one when required to meet specific functional requirements.
- Execution management manages long-lasting interactions and handles changes in resource availability at runtime.

The rendezvous component has the central role in the system. It provides a repository for all resource descriptions currently available in the system as well as two interfaces: publishing and lookup. The publishing interface is used by resources to register in the rendezvous and publish own description. The lookup interface is used by resource clients to search for resources capable of fulfilling the clients' requests. This type of architecture allows late binding of resources, i.e., the applications do not have to define which resources shall be used, but only the type of information required by the application. Based on the description of the required information, the Resource layer provides the actual resources that can most efficiently at a given time provide the requested information.

All resources provide access policies as part of their descriptions, outlining who and under which circumstances can access a given resource. The rendezvous component uses these policies to grant or refuse access to the resources based on their credentials.

The proposed architecture remains to be tested and validated in a test bed combining a number of different applications in a real-world setting. Based on this evaluation the architecture shall be improved. A number of interesting points remain to be resolved and proved like the scalability of the system, interaction between different SENSEI system, management of such system and its components, etc. The project is very well embedded in the EU's FIA initiative and to a great extent influences the design of a Future Internet with the specific real-world requirements.

5.4 Conclusions

In recent years many SNIFs have been proposed as a result of a tremendous increase of heterogeneous WSNs deployments. This chapter describes the existing SNIFs from a historical perspective and compares their functional and nonfunctional properties. Also a description of currently developed systems is provided with a perspective view on trends in mobile and future internet.

Currently there is no existing standard technology for the SNIFs and none of the developed systems has gained prevailing attention as a reference model for future research. However, some of the described frameworks are built by a large consortium of industrial or academic partners (e.g., SENSEI) or are promoted by a standardized organization (e.g., Sensor Web Enablement). Others are developed and available as an open source project used by an increasing number of contributors and users (e.g., GSN). It will take some time until the SNIFs gain broad attention as an integral part of the communication system. Before then, WSNs have to be largely deployed and ubiquitous so that the benefits of SNIFs become pronounced and they will emerge as the only solution to handle and manage the amount of data produced by the sensors. This may happen soon because the WSNs market will grow over eleven times within next 10 years [22]. Then the SNIF which is most advanced and provides broad scope of features will have the biggest advantage and potentially will dominate other solutions.

References

1. Mainwaring A, Polastre J, Szewczyk R, Culler D, Anderson J (2002) Wireless sensor networks for habitat monitoring. ACM International Workshop on Wireless Sensor Networks and Applications, Atlanta, GA, 28 Sept 2002
2. Paek J, Chintalapudi K, Govindan R, Caffrey J, Masri S (2005) A wireless sensor network for structural health monitoring: performance and experience. The Second IEEE Workshop on Embedded Networked Sensors EmNetS-II, Syndey Australia, 30–31 May 2005
3. Lombriser C, Bharatula NB, Roggen D, Tröster G (2007) On-Body Activity Recognition in a Dynamic Sensor Network, In Proceedings of 2nd International Conference on Body Area Networks (BodyNets), Florence, Italy, June 2007
4. Gibbons PB, Karp B, Ke Y, Nath S, Seshan S (Oct–Dec 2003) IrisNet: an architecture for a worldwide sensor web. IEEE Pervasive Comput 2(4):22–33
5. Srivastava M, Hansen M, Burke J. Parker A, Reddy S, Saurabh G, Allman M, Paxson V, Estrin D (April 2006) Wireless urban sensing system. CENS Technical Report #65
6. Dunkels A, Gold R, Marti S, Pears A, Uddenfeldt M (2005) Janus: an architecture for flexible access to sensor networks. In: Proceedings of the 1st ACM workshop on dynamic interconnection of networks DIN '05. ACM, Cologne, Germany, pp 48–52, 2 September 2005
7. Lacoss RT (1987) Distributed mixed sensor aircraft tracking, presented at the American Control Conference, Mineapolis, MN
8. Polastre J, Szewczyk R, Culler D (2005) Telos: enabling ultra-low power wireless research. In: Proceedings of the fourth international conference on information processing in sensor networks: special track on platform tools and design methods for network embedded sensors (IPSN/SPOTS), UCLA, Los Angeles, CA, USA, April 25–27, 2005
9. ZigBee Allince (2008) www.zigbee.org
10. HART Communication Foundation (2009) http://www.hartcomm2.org/index.html
11. International Society of Automation (2010) http://www.isa.org/MSTemplate.cfm?MicrositeID=1134&CommitteeID=6891
12. Gibbons P, Karp B, Nath S, Ke Y, and Seshan S (2003) IrisNet: An Architecture for a Worldwide Sensr Web, In IEEE Pervasive Computing, Special Issue on Sensor and Actuator Networks, IEEE Press, October-December, 2003
13. Shneidman J, Pietzuch P, Ledlie J, Roussopoulos M, Seltzer M, Welsh M (2004) Hourglass: an infrastructure for connecting sensor networks and applications, Harvard technical report TR-21-04
14. Aman K, Suman N, Jie L, Feng Z (Oct–Dec 2007) SenseWeb: an infrastructure for shared sensing. IEEE Multimedia 14(4):8–13
15. Chatzigiannakis I, Mylonas G, Nikoletseas S (2007) The Design of an Environment for Monitoring and Controlling Remote Sensor Networks. Int. J. Distrib. Sen. Netw. 5, 3 (July 2009), 262–282
16. Bernat J, Pérez S, González A, Sorribas R, Villarrubia L, Hernández L (June 2008) Ubiquitous sensor networks in IMS: an ambient intelligence telco platform. ICT Mobile Summit
17. Gluhak A, Schott W (2007) A WSN system architecture to capture context information for beyond 3g communication systems. In: In Proceedings of the third international conference on intelligent sensor, sensor networks and information processing (ISSNIP) 2007, Melbourne Australia, 3–6 Dec 2007
18. Aberer K, Hauswirth M, Salehi A (2007) Infrastructure for data processing in large-scale interconnected sensor network, In Proceedings of 8th International Conference on Mobile Data Management (MDM), Mannheim, Germany, May 2007
19. Botts M, Percivall G, Reed C, Davidson J (eds) (2007) OGC sensor web enablement: overview and high level architecture, White Paper Version 3, Open geospatial consortium Inc., 27 Dec 2007
20. CoBIs Final Project Report (Mar 2007) Deliverable D104, Version 2.0. http://www.cobis-online.de/files/Deliverable_D104V2.pdf. Last visited 2 Sept 2008

21. Krco S, Johansson M, Tsiatsis V (2007) A commonsense approach to real-world global sensing. In: Proceedings of the SenseID: convergence of RFID and wirelesssensor networks and their applications workshop. ACMSenSys 2007, Sydney, Australia, Nov 2007
22. Harrop P, Das R (2008) Active RFID and sensor networks 2008–2018. IDTechEx research report, Feb 2008
23. ITU (2007) TSTAG: A preliminary study on the Ubiquitous Sensor Networks. TSAG-C 22-E. Feb 2007
24. Marin-Perianu M, Meratnia N, Havinga P, Moreira Sa de Souza L, Muller J, Spiess P, Haller S, Riedel T, Decker C, Stromberg G (Dec 2007) Decentralized enterprise systems: a multiplatform wireless sensor network approach. IEEE Wireless Commun 14(6):57–66

Chapter 6
Mobility Aspects in WSN

Aleksandra Mateska, Liljana Gavrilovska, and Sotiris Nikoletseas

6.1 Introduction

Wireless Sensor Network (WSN) is a collection of sensor nodes able to sense their environment, collect and process various data, and communicate among each other. The introduction of mobility of sensor nodes can significantly affect and improve the overall network performances. Sensor nodes may move by self-driving (e.g. mounted on wheels) or by being attached to transporting devices (e.g. robots, people, vehicles, or animals) resulting in longer network lifetime, better area coverage, and dynamic adaptation to different system functionalities and requirements.

Mobile WSNs may include *sensor nodes*, *actuators*, *relays*, and *sinks* (see Fig. 6.1). Actuators are nodes that may dynamically act upon received information, both on sensors and on environment. Relays usually forward the information from the sensors to the so called sink (a data collector and a possible gateway to the backbone network).

The rapid progress of wireless communication and distributed embedded sensor technology has led to the development of many applications of mobile sensor networks. The various applications range from natural ecosystem to security monitoring, particularly in inaccessible terrains or disaster relief operations [1].

This chapter presents a classification of mobility models based on the mobile entities the models are applied on. Then it presents detailed description of sensor mobility and its impact on coverage and network topology, as well as its possible impact for sensing enhancement. In addition, existing data dissemination protocols in WSNs with

A. Mateska (✉)
Aleksandra Mateska, Faculty of Electrical Engineering and Information Technologies, Skopje
e-mail: amateska@feit.ukim.edu.mk

L. Gavrilovska
Faculty of EE and IT, Ss. Cyril and Methodius University of Skopje, Macedonia
e-mail: liljana@feit.ukim.edu

S. Nikoletseas
Research Academic Computer Technology Institute (CTI) and University of Patras, Patras, Greece
e-mail: nikole@cti.gr

L. Gavrilovska et al. (eds.), *Application and Multidisciplinary Aspects of Wireless Sensor Networks*, Computer Communications and Networks,
DOI 10.1007/978-1-84996-510-1_6, © Springer-Verlag London Limited 2011

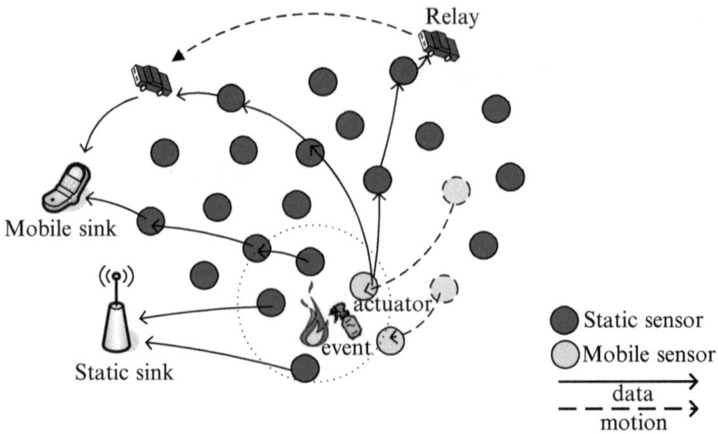

Fig. 6.1 WSN, mobile entities and possible reaction to events

mobile sensor nodes are summarized. Cooperative communications are further discussed as an efficient way for exploiting multi-user diversity for data propagation.

6.2 Mobile Entities in WSN

Each sensor node in the WSNs comprises *sensing unit*, *transceiver*, *processing unit,* and *power supply* (usually battery). If mobile, sensor nodes can be equipped with various locomotion devices and, additionally, they may contain different position finding systems (Fig. 6.2).

The sensor's sensing range defines the *coverage*, while the transmission and reception range of transmitter defines the *connectivity* of the sensor network. It is significant to provide connectivity and coverage at the same time since a sensed data is not beneficial if it cannot be sent to the sink due to poor connectivity [1]. These aspects are also addressed in [2, 3]. The authors in [2] evaluate the minimum transmission range needed for connectivity in static WSNs modeled by random geometric graphs, while [3] provides necessary and sufficient relation between the transmission range and the sensing range to simultaneously guarantee coverage and connectivity.

Sensor devices run on batteries, which makes the *power consumption* another key factor in the design of nodes and network protocols. This introduces another important performance parameter of WSNs, i.e. *network lifetime*. Power consumption affects all aspects of WSNs, from choosing hardware components to programming efficient network protocols for minimizing the use of the radio, and efficient power control algorithms that will switch off hardware components when necessary. Another aspect is the limited computation capabilities of the microcontrollers and the reduced memory space to run the programs. These limitations restrict the protocol stacks and computation algorithms requiring the programming of the sensor devices to be hardware specific [4].

Coverage, *connectivity,* and *network lifetime* are the most important performance metrics that can be improved by enabling the sensor nodes with motion capability.

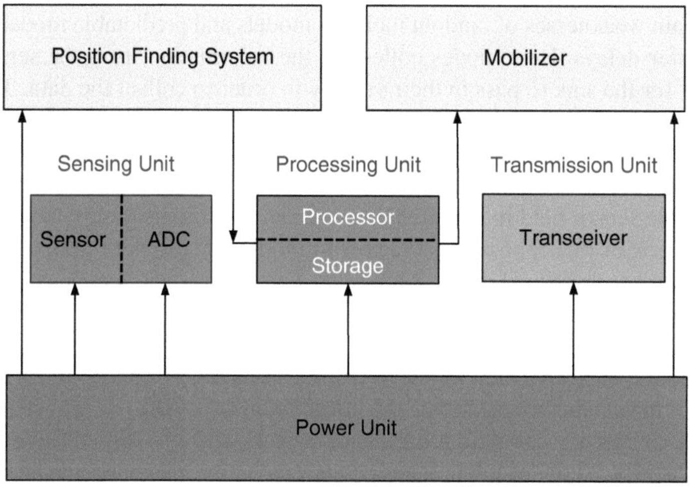

Fig. 6.2 Components of a sensor node

Mobile entities (i.e. sensors, sinks, and/or relays) can follow diverse mobility patterns which lead to different performances improvement. However, sensor mobility, despite the numerous advantages it brings, imposes many challenges when designing appropriate algorithms and protocols. Data dissemination in a mobile scenario is another challenging issue that needs to be properly solved. It can be realized with flooding and direct transmission, which are simple, but usually inefficient. Furthermore, many protocols designed for static networks need to be radically redesigned in order to meet mobility constraints.

The mobile entities of interest in WSN can be *sinks*, *relays,* or *sensors*. Sink mobility may prevent the sink area to become a bottleneck [5]. Mobile relays can inherit other sensors to improve network connectivity. Sensor mobility can be further exploited to compensate for the lack of sensors and improve the network coverage and detection time, which optimize energy utilization and prolong network lifetime.

6.2.1 Mobile Sink Approach

The *mobile sink* approach is a common solution exploiting sink nodes' mobility in order to get closer in the vicinity of the reporting sensors. It also achieves higher degree of *load balancing* among sensor network nodes and can offer extensive improvements of the *network lifetime* [6]. The mobile sink can follow random [7–9], predictable [10], or controlled mobility [11, 12]. Random movement of the sink node can be equivalent with the scenario where people wearing sinks randomly move and collect information from sensors deployed in a certain area (market for example). Predictable mobility can be the movement of a bus or train, while robots can be enabled to achieve specific tasks with controllable movements.

The main weaknesses of random mobility models and predictable models are the data transfer delays. Here, nodes collecting the information from the sensor field must wait for the sink to pass in their vicinity in order to collect the data. This time can have high value causing long delays in the data transfer and making this solution inappropriate for real-time applications. However, if controllable mobility is used, then specific algorithms can guide the movement entities to collect information from the sensor field in a desirable fashion and overcome the problem of delay. The drawback of this approach is the cost of implementation in terms of hardware resources and navigational needs. Moreover, a stronger model (such as global network knowledge) is usually assumed, in contrast to the much weaker assumption of random mobility.

The mobile sink approach has several disadvantages. All nodes must know the position of the sink in order to route the information to it. Also, most scenarios have the sink acting as a gateway to a backbone network and it is difficult to engineer a system whereby a mobile sink is always connected to the backbone network. Mobile sinks perform advantageous regarding network lifetime improvement compared to other mobile entities. In such a way, moving the sink nodes demonstrates better performances than the mobile relay approach, but relays are beneficial in application scenarios where a mobile sink is not feasible (for example, in hostile terrains) [6].

6.2.2 Mobile Relay Approach

Another approach to enhance the connectivity and the network lifetime in WSNs is the *mobile relay* approach. It can lead to *elongating lifetime of bottleneck sensor nodes*. Mobile relay can follow a random, predictable, or deterministic mobility pattern. Randomly moving Data MULEs are introduced as forwarding agents to help gather the sensing data, thereby saving energy due to single-hop transmissions (i.e., from a sensor to a mule that is passing by) [13]. Controllably moving Message Ferry is a mobility-assisted approach, which utilizes a set of special mobile nodes called message ferries to provide communication service for nodes in the deployment area [14, 15]. These two approaches are introduced for sparse sensor networks, while deterministic mobility is suitable for dense sensor networks where the relay follows a deterministic path for achieving network lifetime improvement [6].

Mobile relay approach has serious latency drawbacks leading to a decrease in the achievable throughput. Therefore, this kind of solutions are mostly suitable where mobile sinks cannot be implemented and are robust, flexible, and easier to design [6].

6.2.3 Mobile Sensor Approach

The movements of sinks and relays mainly improve network lifetime and provide more efficient energy utilization. The *mobile sensors*, through *network topology* adjustment, can improve different network performances (i.e., *coverage*, *sensing* etc.). In order to best fulfill its designated surveillance tasks, a sensor network must maximally or fully cover the observed region where the events occur, without internal sensing holes. This task can be accomplished by moving the sensor nodes toward desirable positions and is particularly relevant when an incremental deployment of sensors is not possible. Due to the limited power availability at each sensor, energy consumption will be the primary issue in the design of any protocol for mobile sensors. Since sensor movements and, to a minor extent, message transfers, are energy-consuming activities, the protocols should minimize movements and message exchanges, while following a satisfactory coverage [16].

The main groups of mobile entities were introduced together with their characteristics and targeted network performances. The following subchapter presents the research on recently proposed sensor mobility regarding the design trends for the specific mobility models.

6.3 Design Trends for Sensor Mobility

The design of sensor mobility models faces two important research trends, i.e., approaches for *network topology* and *coverage improvement* and for *sensing improvement*. Connectivity is taken into consideration in some of the approaches as well, as maintaining coverage while satisfying connectivity constraints. Representative examples of the former trend comprise Vector, Voroni, and Quorum-based approaches, and diverse protocols based on different grid structures. The latter issue for sensing improvement takes into account sensor movement approaches toward events that occur in the network.

Ref. [17] considers two basic sensor movement techniques: *sensor self-deployment* and *sensor relocation*. Sensor self-deployment is performed after the initial sensor dropping, while the relocation takes place after discovering the failure nodes. Replacement discovery and replacement migration are the two basic steps considered in the sensor relocation phase.

The following paragraphs discuss sensor mobility approaches classified according to the performance metrics they tend to improve (coverage and sensing). Description of the models and comparison of their advantages vs. disadvantages are also presented.

6.3.1 Coverage Improvement

The proper network coverage is of crucial importance for the WSN operation and reliability. The coverage in a WSN is mainly defined by the sensing ranges of the deployed sensors. Sensors and sinks need to be suitably positioned in the area of interest, thereby enabling the nodes with mobility to achieve this goal. Sensor mobility models that aim to improve the coverage of the WSN can be broadly classified as:

- Vector-based approaches
- Voroni-based approaches
- Quorum-based approaches
- Grid structure approaches

Representative models for each of the enumerated groups are further described.

6.3.1.1 Vector-Based Approaches

Vector-based approaches are extensively used for sensor self-deployment and coverage improvement. The literature presents many variations of the basic technique introduced in [18], where the sensor nodes are treated as virtual particles subject to virtual forces. This approach is originally proposed for mobile autonomous robots and is based on potential fields, which are assumed to exist in the sensor field.

Similar technique is used in [19], where a feature for providing constrained coverage in the sensor field (at least *k*-coverage) is added. The so-called *Potential Field Algorithm* (PFA) works by abstracting the sensor node to be a particle in the potential field, which will be influenced by the forces on the nearby nodes. The force which acts on the sensors is a gradient of the scalar potential field *U*, and is presented with $F = -\nabla U$. The forces between the nodes are obtainable by attractive and repulsive patterns – F_{attr} and F_{rep}, and the resultant force between any two nodes *i* and *j* is given by: $F_{i,j} = F_{attr_{i,j}} + F_{rep_{i,j}}$, where

$$F_{attr_{i,j}} = \left(\frac{-K_{attr}}{\Delta x_{ij}^2}\right)\left(\frac{x_i - x_j}{\Delta x_{ij}}\right) \tag{6.1}$$

$$F_{rep_{i,j}} = \begin{cases} \dfrac{-K_{rep}}{\Delta x_{ij}^2}\left(\dfrac{x_i - x_j}{\Delta x_{ij}}\right), & \text{if critical connection} \\ 0 & \end{cases} \tag{6.2}$$

and x_i is the position of the *i*-th node, Δx_{ij} is the Euclidean distance between nodes *i* and *j* and K_{attr} and K_{rep} are the force constraints. The attractive and repulsive forces follow inverse square law depending on the distance between the sensors. The attractive forces – $\|F_{attr}\|$ tend to infinity when the distance between the nodes is zero thus avoiding collisions, while the repulsive forces – $\|F_{rep}\|$ tend to infinity when the distance between the neighboring sensors is equal to the communication radius, thus preventing the nodes to lose connectivity. The force that will act on the *i*-th node is $F_i = \sum_{\text{neighbors}, j \neq i} F_{i,j}$ and the node will move according to $\ddot{x}_i = (F_i - \nu \dot{x}_i / m)$, where ν is a chosen dumping factor and *m* is the virtual mass of the node (assumed to be one).

The presented sensor self-deployment approach is advantageous because it allows nodes to use only their local knowledge to make their deployment decision. It does not require centralized control and localization. Possible weaknesses which can be enumerated include coverage holes near the obstacles and frequent topology changes (if nodes failures are often, thus causing frequent requirements for nodes movements). It is computationally expensive and assumes that each node can sense the exact relative range and bearing of its neighbors.

A similar idea with a different approach is presented in [20], where the sensor locations are determined based on a virtual force assumed to exist among sensors and between the sensors and obstacles in the field. Each sensor locally calculates the sum of the vectors (representing relative positions) of neighboring nodes acting it and moves according to the movement vector thus computed.

In the so-called *virtual force*–based mobile sensor deployment algorithm (VFA), there is a powerful cluster head (CH) which executes the algorithm. The CH communicates with all the sensors, collects sensor position information, and calculates forces and desired position for each sensor. In VFA, the distance between two adjacent nodes, when all nodes are evenly distributed, is defined as a threshold to distinguish attractive or repulsive force between the nodes. The total force on a node will be the sum of all the forces given by other sensors together with obstacles and preferential coverage in the area (Fig. 6.3, Eq. 6.3). In terms of forces, the sensor nodes are subject to three types of forces: repulsive forces due to the obstacles (F_{rep}), attractive forces from the areas with pure coverage (F_{attr}), and forces (either attractive or repulsive) from the other sensor nodes depending on their distance and orientation (F_{sensor}).

$$F_i = \sum_{j=1, j\neq i}^{k} \left(F_{sensor_{ij}}\right) + F_{rep_i} + F_{attr_i} \tag{6.3}$$

Among the virtual force and potential field based approaches, a *distributed self-deployment algorithm* (DSSA) has been proposed in [21]. The objective of

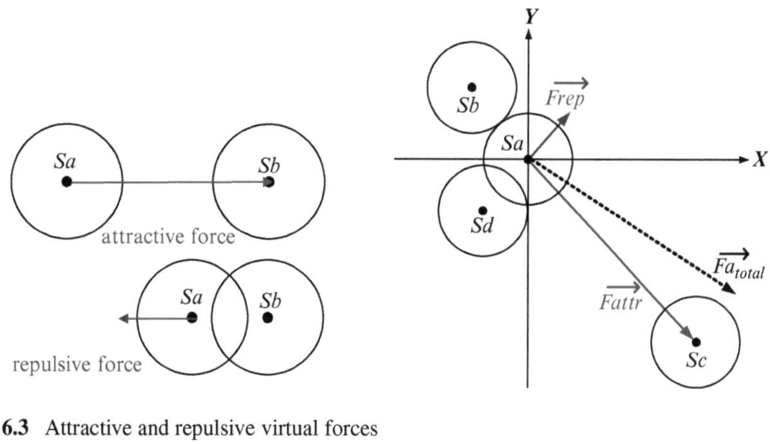

Fig. 6.3 Attractive and repulsive virtual forces

DSSA is to maximize coverage and to maintain uniformity of node distribution. Similar to VFA, it uses the concept of virtual force that depends on the internodes separation distance and local current density (D_n). In the beginning of the algorithm, the initial density for each node is equal to the number of its neighbors. The algorithm defines the concept of expected density (μ) as the average number of nodes required to cover the entire area, when the nodes are deployed uniformly. Depending on the forces from the neighboring nodes, a node can decide on its next movement location. The algorithm works iteratively, and the force of the i-th node as a result of the j-th node in the n-th step is:

$$f_n^{i,j} = \frac{D_n^i}{\mu^2}\left(cR\left|p_n^i - p_n^j\right|\right)\frac{p_n^j - p_n^i}{\left|p_n^j - p_n^i\right|} \tag{6.4}$$

where cR is the communication range of the sensor, and p_n^i is location of the i-th node in the time stamp n. The algorithm terminates when a node moves an infinitely small distance over a period of time or when it moves back and forth between two same locations (performing stability and oscillation check). DSSA brings the advantages over the previous two approaches in overcoming the problem of possible oscillatory sensor behavior. But, it does not consider the possible obstacles, and assumes that each node knows its location.

One of the techniques that will help terminate the nodes movement in a desirable fashion (achieving coverage effectiveness) is a Voronoi diagram approach, introduced in [22]. Here, the nodes move toward coverage holes in the network, thus providing high coverage with short deploying time and limited movement from densely to sparsely deployed areas.

6.3.1.2 Voroni Approaches

Many research groups utilize the idea of partitioning the area of interest into Voroni polygons when building algorithms for sensors self-deployment and relocation. The Voronoi diagram of a collection of nodes partitions the space into polygons, where every node (sensor) is enclosed in its own polygon. Every point in a given polygon is closer to the node in its polygon than the sensors positioned elsewhere. Thus, if a sensor cannot detect the expected phenomenon, no other sensor can detect it, and then each sensor is responsible for the sensing task in its Voronoi polygon. In this way, each sensor can examine the coverage hole locally (Fig. 6.4).

The objective of the approaches for sensor movements, which use Voroni diagrams, is to minimize the sensors' local uncovered areas. This is executed by iteratively aligning their Voronoi regions with their sensing range. The Voronoi-based algorithms differ in their arrangement methods. In [21], a node moves to the point that maximizes a utility metric defined as the product of the node's effective area and the node's estimated lifetime, while in [22], the best possible improvements are achieved when the nodes move half of the communication range toward their

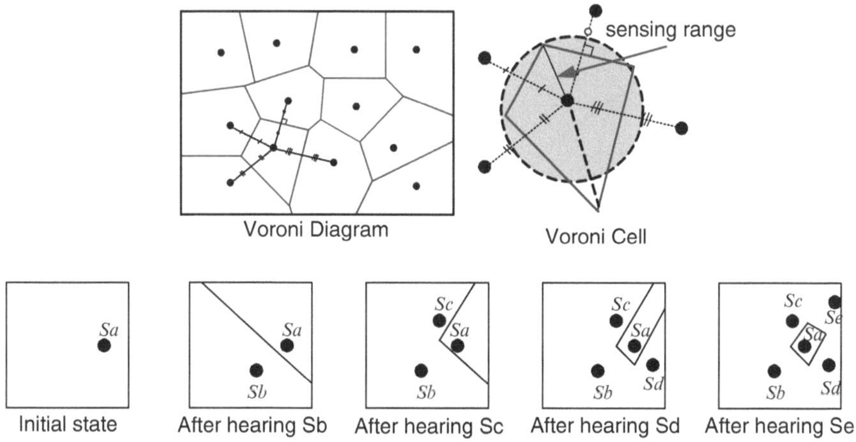

Fig. 6.4 Voroni Diagrams, Voroni Cell, constructing Vorioni diagrams

furthest Voronoi vertexes [17]. Voroni diagrams need to be repeatedly constructed to reflect the nodal movement. Since the construction of the diagram requires global computation, this approach has large message overhead. To avoid oscillations (e.g. moving back and forth between several points), nodes may stop moving early which can cause coverage redundancy and coverage holes in the network.

The following paragraphs present approaches based on a Voroni-diagram and "quorum-based" approaches, which are used for coverage improvement in terms of efficiently covering the *coverage holes*.

Wang et al. [22] describe three distributed self-deployment algorithms (VEC, VOR and min–max) for mobile sensors using Voroni diagrams. Once the Voroni polygons are constructed, each sensor within the polygon can examine the existence of possible coverage holes. If such a hole is discovered, the sensors will move to new positions according VEC, VOR, or min–max protocol.

The *Vector-based algorithm* (VEC) pushes sensors from densely to sparsely covered areas. Two sensors exert a repulsive force when they are close to each other. If d_{av} is the average distance between any two sensors, the virtual force between the sensors s_i and s_j will move each of them $\left(d_{av} - d\left(s_i, s_j\right)\right)/2$ distance away from each other. In case one of the sensor's sensing range completely covers its Voroni polygon, only the other sensor should move away $\left(d_{av} - d\left(s_i, s_j\right)\right)$ distance. In addition to the mutual repulsive forces between sensors, the boundaries also apply forces to push sensors inside the boundary. If $d_b\left(s_i\right)$ is the distance of a sensor s_i from its closest boundary, then the repulsive force would move it a distance $d_{av}/2 - d_b\left(s_i\right)$ opposite the boundary.The *Voroni-based algorithm* (VOR) pulls sensors toward their local maximum coverage holes. If a sensor detects a coverage hole within its Voroni polygon, it will move toward its farthest Voroni vertex, such that the distance from its new location to its farthest Voroni vertex (*v*–*u* in the Fig. 6.5) will be equal to the sensor's sensing range. This way, the maximum moving distance for a sensor is limited to at most half the communication range.

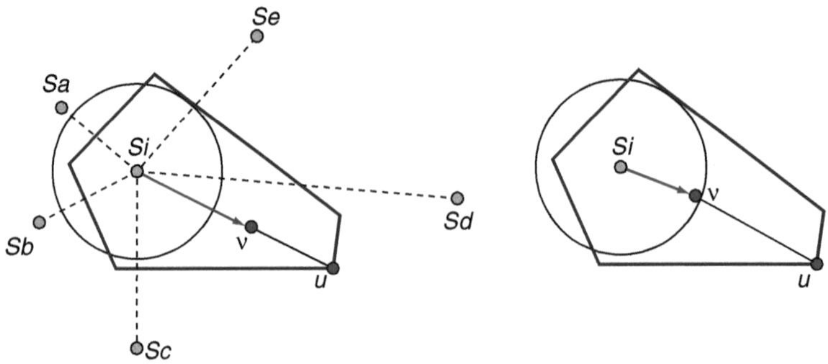

Fig. 6.5 VOR and min–max strategy

The *min–max algorithm* is very similar to VOR, but it moves a sensor inside its Voroni polygon to a point where the distance from its farthest Voroni vertex is minimized. Since moving to the farthest Voroni vertex might lead to a situation for the vertex that was originally close to becoming a new farthest vertex, the algorithm positions each sensor in a manner that no vertex is too far away from the sensor.

Utilizing local movements and communications, the proposed protocols are extensible for large deployment scenarios, but they perform better if initial deployment is random rather than clustered and does not perform well in case of insufficient communication range.

Bidding and *proxy*-based sensor deployment are protocols proposed by the authors in [22], which are based on Voroni diagrams and are created for handling coverage holes in a hybrid network structure composed of static and mobile nodes.

In [23], each static sensor is location aware and can detect a possible hole, based on the constructed Voroni diagram. Mobile sensors are used to cover the holes after the static sensors, which have discovered a hole, have sent requests (*bids*) to the nearest mobile sensors. If a hole is found, a static sensor chooses the location of the farthest Voroni vertex as the target location of the mobile sensor and calculates the bid as $\pi\left(d-R_s\right)^2$, where d is the distance between the sensor and the farthest Voroni vertex and R_s is the sensing range. A static sensor then finds the closest mobile sensor whose base price (each mobile sensor has an associated base price that is initialized to zero) is lower than its bid and sends a bidding message to this mobile sensor. The mobile sensor receives all such bids from its neighboring static sensors and chooses the highest bid and moves to heal that coverage hole. The accepted bid becomes the mobile sensor's new base price. Possible disadvantages of the protocol are energy consumption when nodes move necessarily (applying iterative movements) and the possibility of moving excessive number of nodes toward the same hole (duplicate healing problem).

To overcome the disadvantages of the bidding protocol, and additionally to handle with multiple holes, the authors have proposed *proxy-based sensor relocation protocol* in [24]. Instead of moving iteratively (Fig. 6.6a), the sensors calculate their target location in each step while moving logically and exchanging their logical information with the logical neighbor nodes. The movement occurs only after the

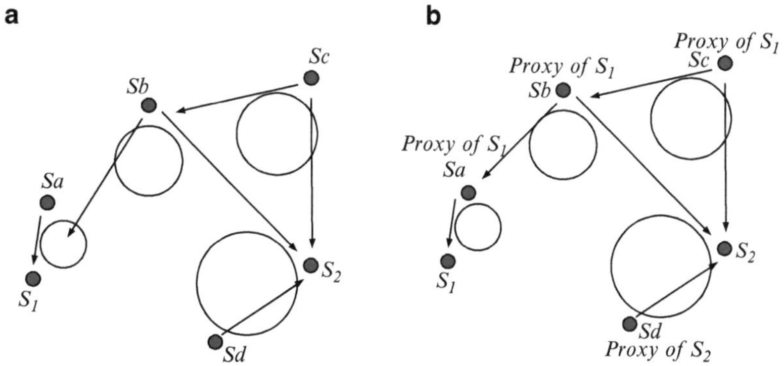

Fig. 6.6 (**a**) Iterative sensors movement and (**b**) proxy based approach

algorithm has calculated its target location. Static sensors construct a Voroni diagram and listen to the mobile sensors' advertisements. After receiving an advertisement, a static node records the embedded information and maintains a mobile sensor list. Once a static sensor detects a sensing hole in its Voroni polygon, it estimates the hole size and computes a bid accordingly. Then, from its mobile sensor list, it chooses the closest one with the lowest base price that is smaller than the bid, and sends a bidding message to that sensor. In the case that a mobile sensor receives more than one bidding message from different static sensors, it chooses the highest bid and sends a delegate message to the corresponding bidder. After receiving the delegate message, the bidder becomes the proxy of the mobile sensor and executes the relocation protocol on its behalf as if the mobile sensor had migrated to the sensing hole. The approach addresses the issue of minimization of energy consumption by letting the nodes move locally, but only after they have decided whether their movement maximizes the coverage or not (Fig. 6.6b). Even though the main advantages of the proposed protocol are avoiding the multiple holes and thus load balancing, it is complex for implementation and requires cooperative computing of the sensors to determine their logical movement. The protocol is bandwidth inefficient because of the frequent advertisement broadcasting. Additionally, both bidding and proxy employ moving toward largest coverage holes.

6.3.1.3 Quorum-Based Approaches

Several *quorum-based* approaches are introduced in literature [22, 25, 26]. Ref. [22] presents the original quorum-based approach, where filling the coverage holes with cascade sensor movements is performed from cells with redundant sensors toward cells with a lack of sensors. The protocol partitions the area in grids and assigns a CH or responsible node in each cell. The CH will process the advertisements and queries from the cells with redundancy and deficiency of sensors, respectively (Fig. 6.7a). This approach is described in more detail in the following section, where the approaches which use different grid structures are classified.

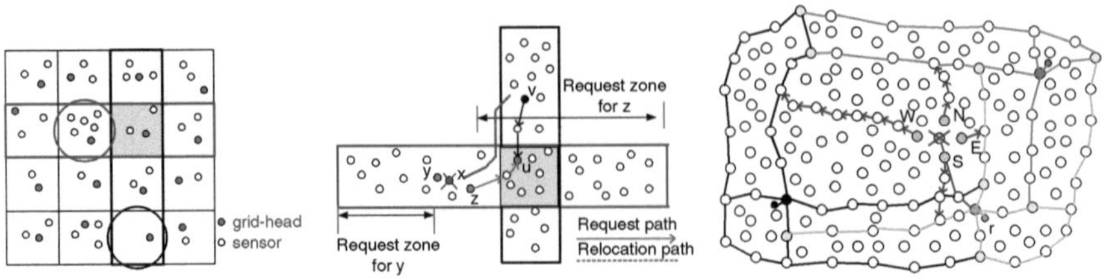

Fig. 6.7 A general view of how (**a**) quorum based (**b**) ZONER (**c**) MSRP approach works

Variant of the quorum-based location service is the localized sensor relocation protocols *ZONER (*zone-based sensor relocation protocol), proposed in [25]. In ZONER, redundant sensors distribute their location information within a vertical registration zone. After a node failure, two specified neighbors (on both the right and left sides) query sensors within their bounded horizontal request zones for the nearest redundant node. Thus, the discovered redundant node is relocated by shifted movement (as replacement to the position of the failed node along a natural migration path). Since the algorithm does not use network-wide flooding, and because only selected nodes are required to move, ZONER is both bandwidth and energy efficient (Fig. 6.7b).

In mesh-based sensor relocation protocol (*MSRP*), proposed in [26], redundant sensors choose the nearest non-redundant sensors to act as a proxy node running the iMesh algorithm (distance-sensitive service discovery algorithm) on their behalf to construct an information mesh. When a node fails, the four neighbors search in four directions to discover nearby proxies. The nearest detected redundant sensor of the closest discovered proxy node is then relocated (by shifted movements) to replace the failed node along an energy-aware migration path (Fig. 6.7c).

According to the authors, both ZONER and MSRP outperform the existing sensor relocation protocols (the quorum- and proxy-based approaches) because of their localized message transmission, zero requirement of pre-knowledge of the network, and guaranteed node replacement. Moreover, MSRP is also better than similar algorithms (including ZONER) that rely on shifted replacement migration for their novel localized migration path discovery.

6.3.1.4 Grid Structure Approaches

Different approaches for sensor self-deployment and relocation use various types of *grid structures* on top of the physical infrastructure. In this manner, appropriate basis for sensor movement algorithms application can be defined. The cost of organizing sensors into grids is low, and it is an easy method for facilitation of data aggregation and routing.

The existing literature offers approaches where the sensor area is divided into two dimensional evenly partitioned grid structures [27, 28], or in hexagonal grids [1, 29]. A novel approach using triangle tessellation (TT) layout is derived in [30].

Approaches with Square Grid Structure

In [27], an algorithm called *SMART* (scan-based movement-assisted sensor deployment method) is introduced, which partitions the region of interest in a 2D mesh through clustering. The algorithm is distributed and scan based, nodes are treated as a load and the objective is to balance the load (number of nodes) in each cell. Each cluster corresponds to a square region and has a CH that is in charge and which communicates with adjacent CHs. A hybrid approach is used for load balancing, where the 2D mesh is partitioned into 1D arrays by row and by column. Two scans are used in sequence: one for all the rows, followed by the other for all the columns. Within each row and column, the scan operation is used to calculate the average load and then to determine the amount of overload and underload in clusters. Load is shifted from overloaded clusters to underloaded clusters to achieve a balanced state. In areas with holes, a preprocessing is performed for planting "seeds" in holes at each 1D scan. These seeds will serve as CHs in the holes. This approach requires the network to be dense enough so that load balancing can be proceeded in the entire sensory field, what may generate huge message overhead. A simple example of a 2D scan is presented in Fig. 6.8.

A grid-quorum approach is considered in [29], where the sensor field is evenly partitioned into grids, and each cell has a cell head (Fig. 6.7a). The cells with redundant nodes advertise to other cells in a row, while the cells that need redundant sensors send queries to cells in each column. Since there must be an intersection cell between each row and each column, the intersection cell head will be able to serve the query. To reduce message complexity, information about already discovered closest redundant node is piggybacked on the search message and used to restrict the distance that the message may travel further. Having obtained the location of the redundant sensor, it is moved towards the destination by cascaded movements. Moving it directly to the destination can be a possible solution, but it may take a longer time and it can consume too much energy. With the proposed relocation algorithm, sensor relocation time is reduced. Although the total moving distance may increase, each mobile node moves much less to balance the energy consumption and hence increase the network lifetime.

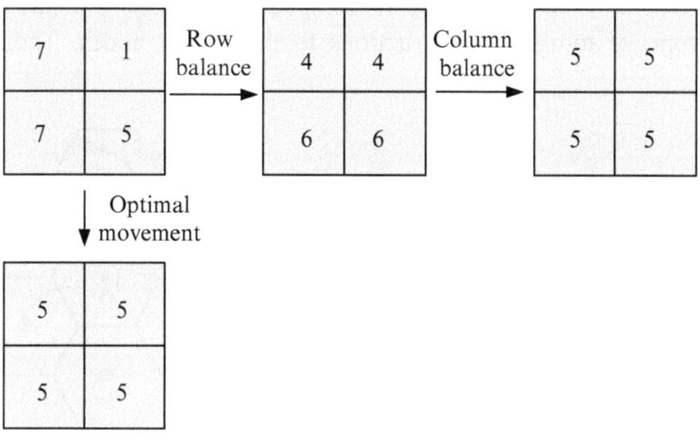

Fig. 6.8 2D SMART algorithm

Approaches with Hexagonal Grid Structure

Partitioning the sensor field in hexagonal grid structure is considered in [1, 16]. *SNAP & SPREAD* is an algorithm for mobile sensor self-deployment, which uses network self-configuration and self-adaptation [16]. The protocol constructs hexagonal grid structure and arranges sensors at centers of thus constructed hexagons. The hexagon edge length is equal to R_s (sensing range) and the proposed deployment strategy guarantees coverage when $R_s \leq \sqrt{3}R_{tx}$, and R_{tx} is the transmitting range of the sensors. In this algorithm, each mobile sensor spontaneously starts to construct a hexagonal tiling; it chooses its current position as the center of the first hexagon of the tiling and becomes snapped. A snapped sensor learns the status of its neighbors, and selects at most six of the neighbors, and snaps them to the centers of adjacent empty hexagons. The new snapped sensors continue with an analogous process until no other snaps are possible. After the snap activity, if there are still spare sensors, a spread process is established where un-spread sensors are pushed toward low-density zones (Fig. 6.9). Each sensor needs locally available information to decide about its movements. The proposed algorithm quickly converges to a uniform and regular sensor deployment over the region of interest, independently of its shape and of the initial sensor deployment. It makes the sensors traverse small distances, avoiding useless movements, and thus the algorithm ensures low-energy consumption and stability. The algorithm can cause network partitioning through constructing separate clusters in sparse networks.

A deterministic approach for deploying sensor nodes into the sensor field is proposed in [1], where hexagonal grid positioning is used to address and locate each sensor node. The dynamic relocation algorithm is called *MaxNetLife*, and is mainly based on utilizing the remaining power of individual sensor nodes as well as properly relocating sensor nodes so that all sensor nodes can transmit the data they sense to the sink. The sensor field is divided into clusters, and clusters are further divided into hexagonal cells. Each CH calculates the energy level of the nodes in its cell and sends the report to the sink node. The sink node then determines if there is a requirement for the nodes to move toward certain locations and which nodes are appropriate for the movement. Then the sink sends mitigation instructions to the CHs, which resend appropriate mitigation instructions to the sensor nodes. Then the actual

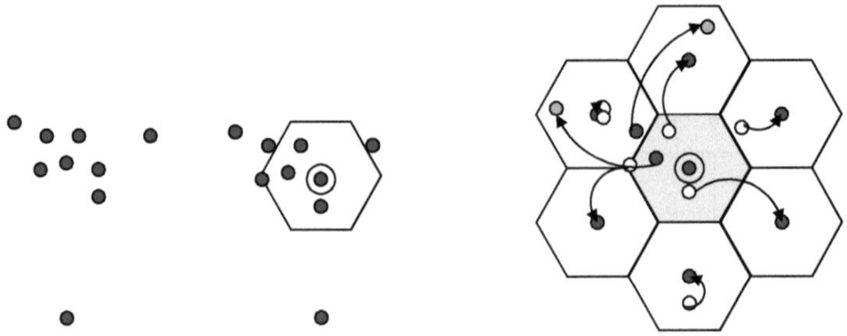

Fig. 6.9 SNAP & SPREAD algorithm

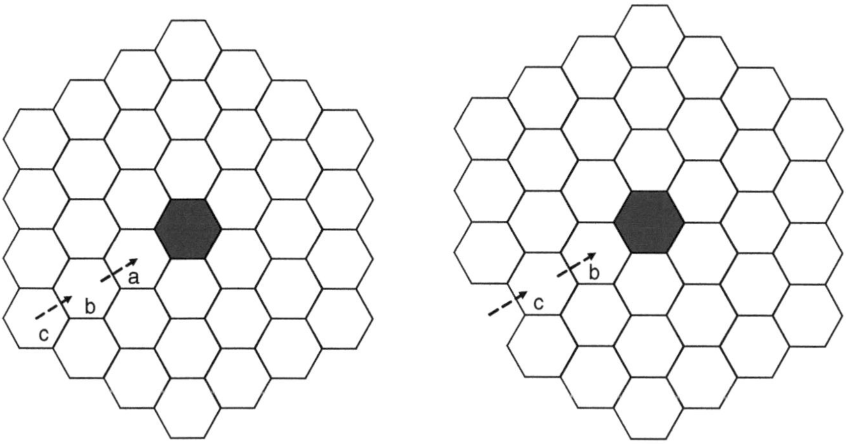

Fig. 6.10 Filing-the-holes with sliding

relocation is performed. The proposed relocation algorithm, of filling-the-hole by sliding model (Fig. 6.10), provides continuous connectivity of the sensor nodes within each cluster by filling out the coverage holes using the sensor nodes within the neighbor cells. When a master node dies in a cell, one of the redundant nodes (if such a node exists) will become a master node. Otherwise, a coverage hole occurs. In this case, a sensor node from the neighbor cell relocates to the coverage hole. If there is a redundant node in the neighbor cell, then that node relocates. Otherwise, the master node in the neighbor cell relocates. Hence, connectivity of all sensor nodes within the cluster is satisfied continuously. Authors use the term *sliding* for filling out the hole in the inner cell by a node from the outer cell. After consecutive sliding relocations happen, a hole may occur in the outermost tier, and hence relocation from other clusters may be required to fill the holes in the outermost tier of the cluster. The algorithm performs locally, and in dense and large networks a long time can be required for the algorithm to terminate.

The TT layout is considered in [30], and two approaches for sensor self-deployment are introduced – *Greedy Advance (GA) and Greedy-Rotation-Greedy (GRG).* An equilateral TT is a planar graph composed of congruent equilateral triangles. It is an appropriate solution because it maximizes the coverage area of any given number of nodes without a coverage gap when the nodal separation equals $\sqrt{3}R_s$ and when the nodes are placed on the vertices of the layout.

The GA moves the nodes along the TT edges toward the given Point of Interest (PoI) where the nodes move to the new vertex closer to the PoI (calculated in the graph distance). Each vertex of TT is represented with three coordinates, and three moving rules are further presented. The first rule is called Priority Rule and determines the priority for the nodes' simultaneous movements to the same vertex. Forbiddance Rule is introduced to avoid simultaneous movement, and the Innermost-Layer Rule allows the three closest nodes to the PoI to move to the PoI, in which case after some node reaches the destination, no other node will be allowed to move to it.

GRG is a combination of GA and a rotation which is applied in a node when its GA movement is blocked. Rotation around PoI is performed in a counterclockwise direction, and a node will stop rotating after it reaches a vertex when greedy movement can resume (or when the rotation in the next hop is occupied, or else when it returns to the starting vertex). The nodal rotation will form the final network in a hexagon shape. The suspension and competition rules can be further implied.

The algorithms are localized self-deployment algorithms, and performance analyses demonstrate that in a finite time they yield to a connected network with full coverage. However, the results of analyses are achieved without taking into consideration the possible obstacles in the sensor field.

Comparisons among the various strategies for coverage improvement are summarized in Table 6.1. The comparison is based on the goals and advantages vs.

Table 6.1 Comparison of different approaches for coverage improvement

Algorithm	Goals and performances	Advantages	Disadvantages
PFA	Coverage improvement while maintaining k-connectivity	Scalability, no need of centralized control and localization	Computation expensive, assumption that every node knows the position and bearing of the other nodes
VFA	Increasing coverage after random sensor deployment	Only local knowledge, no need of centralized control and localization, flexibility, negligible computation time	Possible coverage holes near the obstacles, computation expensive
DSSA	Maximizing coverage and maintaining uniformity	Distributed algorithm, limits possible oscillations	Requires location awareness, obstacles are not considered
VEC, VOR, min–max	Reducing or eliminating coverage holes	Distributed algorithms, local communications and movements	Poor performances in disconnected and clustered networks
BIDP	Coverage improvement with distributed allocation of mobile nodes to the coverage holes	Distributed protocol, cost balance	Several nodes can move towards same hole, requires knowledge of the nodes positions
PROXY	Reducing energy consumption and load balancing, while maintaining coverage	One time movement, multiple healing avoidance	Requires location awareness, bandwidth demanded, high complexity for implementation
Quorum based	Reduction of the sensor relocation time, balance of the energy consumption	Distributed algorithm, relocation towards failed nodes	Requires pre-knowledge of the sensor field, message overheads, replaces sensors after failing

(continued)

Table 6.1 (continued)

Algorithm	Goals and performances	Advantages	Disadvantages
ZONER	Coverage improvement with energy efficient approach for sensor relocation	Guaranties replacement discovery and closest node replacing, void area capability	Local protocol, sometimes increased message overhead
MSRP	Coverage improvement with improved mitigation path discovery	Low message complexity and constant storage load	Local protocol, sometimes can cause increased message overhead
SMART	Even sensor deployment	Distributed protocol, improved convergence rate	Works for dense networks, message overheads
SNAP & SPREAD	Uniform and regular sensor deployment, coverage uniformity	Local knowledge, avoids useless movements, energy effective	Does not guarantee hole elimination, possible network partitioning
MaxNetLife	Increasing cumulative connected coverage	Preemptive relocation of sensor nodes, scalable and applicable	Localized protocol, long termination time
GA & GRG	Connectivity and coverage	Finite execution time	Obstacles are not taken into consideration
EBD	Coverage maintenance, convergence towards event distribution	Distributed, minimal communication and computation	Requires knowledge of nodes positions
SEC	Bounding event loss probability	Good analytical algorithm design	Only analytical analyzes, PoI are known a priori
EDSSA	Jointly achieves network deployment and relocation	No centralized control, no communication between distant nodes	Similar constraints as virtual force based algorithms

disadvantages of the approaches, which were described in the algorithms' definitions. The table gives a list of several algorithms, with their major characteristics and specifics.

6.3.2 *Sensing Enhancement*

Mobile sensors have the advantage to move to locations that meet sensing coverage requirements. Enhancing sensing in WSN can be achieved with repositioning sensors to mitigate sensor failure and moving toward events that occur in the network and that need to be captured in real time. Strategies about the specific problems are further discussed, while their comparisons are added on the previous Table 6.1.

An *event-based deployment* (EBD) framework is introduced in [31], where sensors move to locations where events are detected, while maintaining a desirable coverage for the entire field. The presented approach consists of two distributed schemes for mobile sensors which react to events such that the distribution of the group of sensors tends toward the distribution of the sensed events. The algorithms rely on an initial uniform distribution of sensors and can then generate good results without requiring the sensors to communicate and exchange the information about their positions to each other. One technique uses a very simple update rule to move each sensor based only on the position of the event and the sensor itself (similar to the potential field's approach described in the previous section). This movement is easily computed and the amount of motion can be easily bounded. The second technique uses more computation and keeps a compressed history of the events. Here, to determine its correct position, each sensor maintains a discrete version of the CDF (cumulative distribution) which is updated after each event. The CDF is scaled based on the number of events and the length l of the interval of interest, so that CDF(l) = l. Then each segment of the CDF is associated with a proportional number of sensors, and thus the sensor density will truck the event density. Or in other words, each sensor chooses its new position so that the CDF at this position returns its initial position.

Another method for *stochastic event capturing* (SEC) is presented in [32]. The authors have derived algorithms for motion planning based on defined quality of coverage (QoC) metrics, such as the fraction of events captured and the probability that an event is lost. The analytical results provide guidelines for choosing the velocity and the number of sensors to be deployed for satisfying constraints on fraction of captured events. Only mathematical analyses and proofs derived on approximation factor are presented, while neither simulations nor emulations are made to test the feasibility of the proposed solutions.

Another event-driven sensor self-deployment algorithm (EDSSA) is proposed in [33], based on the virtual force algorithm, where the potential force of the detected event is added to the calculation of the vector sum. This force pulls distant nodes toward the event location and pushes nearby nodes away. By adjusting the event intensity threshold, different node density can be achieved around the event location. The approach considers large-scale relocation of nodes, compared with relocation toward coverage holes in some of the previous approaches. Starting from the initial topology, the nodes will tend to a regular triangular pattern position, achieving the optimal coverage when the length of the grid parts is $D = \sqrt{3}R_s$, where R_s is the sensing range.

These examples for sensor mobility are representative examples of the models found in literature. They form the basics allowing for more comprehensive study later on, since familiarity with the state-of-the-art approaches and their pros & cons is helpful for a research in the considered area. Moreover, the following text presents the main mobility challenges and possible ways to handle them.

6.4 Mobility Challenges: Data Dissemination

Mobility in WSNs is a permanent and challenging issue. It is often followed by network topology modification, necessity of new protocols, and consequently, requirement of appropriate mobility models. Above all, challenges lie in designing cost-efficient routing protocols, which can efficiently disseminate information in the sensor network using resource- and mobility-adaptive algorithms. Therefore, more appropriate and novel algorithms and protocols need to be defined [34].

One of the most prominent research aspects in mobile WSNs is the efficient *data propagation*. Several data dissemination protocols have been proposed for WSNs with static nodes, starting from the basic approaches of flooding and direct transmission, to more efficient transmission protocols.

Flooding technics consume more resources than necessary and should be avoided whenever possible. It is a simple scheme where each sensor aggressively propagates its data messages to any neighboring nodes, resulting in the lowest delivery delay, but causing high overhead and energy consumption. In the direct transmission design, the messages from the sensors are directly transmitted to the sink, and a sensor does not receive or transmit any data messages of other sensors. This is inappropriate in large networks where the sensors can be far away from the sinks. Direct transmission will be too energy consuming or can fail due to the limited communication range of the sensor nodes.

Directed Diffusion (DD) [35] and Low-Energy Adaptive Clustering Hierarchy (LEACH) [36] are widely used data propagation and routing methods. The DD approach assumes that each sink periodically must flood its location information through the sensor field. This procedure sets up a gradient from the sensor nodes to the sink so that each sensor is aware of the sink location for sending future events and measurements. However, such a strategy does not scale with the network size and increases the network congestion, while in a mobile scenario it is an inappropriate solution because of the numerous topology changes. LEACH is a hierarchy-based protocol, where elected CHs transmit aggregated data directly to the sink. It is appropriate only for small networks, and it does not really support movement of nodes. M-LEACH is a proposed protocol which adds a feature to LEACH to support mobile nodes and also reduces the consumption of the network resources [37]. The nodes with less mobility are preferred to be CHs, and the basic idea of the proposed approach is to divide the sensing area into sub-areas and try to optimize the location of CH in these sub-areas.

Overlaying a virtual infrastructure over the physical network has often been investigated as an efficient strategy for effective data dissemination in the presence of mobile sinks. This strategy is implemented in protocols like two-tier data dissemination (TTDD) [38], geographic hash table (GHT) [39], line-based data dissemination (LBDD) [40], and dynamic directed backbone (DDB) [41]. They all use the concept of virtual infrastructure, which acts as a rendezvous area for storing and retrieving the collected measurements. The sensor nodes belonging to the rendezvous area are designated to store the generated measurements during the absence of the sink. After the mobile sink crosses the network, the designated nodes are queried to report the sensory input. A survey of the existing data dissemination protocols for network containing mobile sinks can be found in [42].

A research trend on data dissemination in WSNs where the mobility of some of the sensor nodes is exploited for facilitating the delivery of the sensed data to the sinks, thus enhancing the system's performances, is gaining more attention recently. Data propagation in a scenario with *mobile sensor nodes* is a completely different issue and requires novel approaches since maintaining any kind of infrastructure under these circumstances is impossible. Several approaches considering this topic will be further discussed.

6.4.1 Data Transmission Strategies in WSN with Mobile Sensor Nodes

Two approaches where data dissemination is based on history are initially discussed. Then, a more efficient solution for adaptive data dissemination considering mobility is presented. The last part explains how a cooperative communications can be utilized for efficient messages delivery in a mobile scenario.

A simple and efficient data delivery scheme for special kind of a mobile WSN, called *DFT-MSN* (Delay-Fault Tolerant Mobile Sensor Network), based on delivery probability and fault tolerance is given in [43]. DFT-MSN has several unique characteristics such as sensor mobility, loose connectivity, fault tolerability, delay tolerability, and buffer limit. In the protocol, the decision on when and where to transmit data messages is based on the delivery probability, and the decision which messages to transmit or drop is based on the fault tolerance. The fault tolerance of a message is defined to be the probability that at least one copy of the message is delivered to the sink by other sensors in the network, or according to the hop count of the message. When a message is generated, its fault tolerance is initialized to be zero, so the messages with a smaller fault tolerance are considered more important and are transmitted with a higher priority (by sorting the messages in the queue with an increasing order of their fault tolerance based on their hop count).

A data transmission decision is made based on the delivery probability. At first, the nodes learn their neighbors' delivery probabilities and available buffer spaces via simple handshaking messages. Then, the message is sent to a set of neighbors with higher delivery probabilities. At the same time, the total delivery probability of the considered message is controlled, just enough to reach a predefined threshold, in order to reduce unnecessary transmission overhead. The message transmission scheme is equivalent to direct transmission when the network is just deployed; as the delivery probability is gradually updated with nonzero values, multi-hop transmission will take place. Two data delivery schemes are further proposed, the Replication-Based Efficient Data Delivery Scheme (*RED*) and the Message Fault Tolerance-Based Adaptive Data Delivery Scheme (*FAD*) [44]. The RED scheme utilizes the erasure coding technology in order to achieve the desired data delivery ratio with minimum overhead. It consists of two key components for data transmission and message management. The former makes the decision on when and where to transmit data messages according to the delivery probability. The latter decides the optimal erasure coding parameters (including the number of data blocks and the needed redundancy) based on its current delivery probability. The FAD scheme employs the message fault tolerance, which indicates the importance of the messages.

Another technique for data propagation based on history is proposed in [45]. The considered case study for habitat monitoring assumes that the sensor nodes (zebras in this case) are mobile and the base station is also mobile (sinks mounted on vehicles). Transmitting the data by flooding can provide high success gain, but it requires high bandwidth and capacity, and it's energy consuming. A more efficient protocol is further proposed, where each node is assigned with hierarchy level depending on its success rate on delivering the messages to the base station. A higher level is assigned to nodes that have higher probability of being in range of the base station, based on the nodes' past behavior. When the node has a data which needs to be transmitted, it sends the data to the neighbor with the highest hierarchy level, after it has scanned all the neighbors to acquire the information about their levels. There is a mechanism for decreasing or increasing the hierarchy level in defined intervals, depending on whether the nodes are approaching or leaving the base station. Although the algorithm outperforms flooding in a high-mobility case it can decrease the success rate because of redirecting the traffic toward nodes that are no longer near the base station.

The presented approach in [46] assumes diverse sensor mobility and uses adaptive data dissemination protocols that use mobility level estimation. The approach exploits high mobility as the sensors will dynamically propagate less data in the presence of high mobility, while nodes of high mobility will be favored for moving data around. Furthermore, the message flooding scheme is mobility and progress sensitive, which means that probabilistic forwarding will decrease when the hop count increases and the probability of flooding will decrease when the level of mobility is increased.

The presented protocol assumes that the sensors are mobile (following diverse and time variant mobility) and that they are aware of their position. Novel metric, called mobility level, is introduced – it assigns higher values to nodes that move fast and tend to traverse new areas, while smaller values are assigned to nodes moving slowly and traversing the same or neighboring areas frequently. To increase the probability of data delivery and at the same time reduce delivery delay, data messages are disseminated to several neighbor nodes. If the nodes move slowly, they should choose a greater number of nodes to whom they should transmit the message, and vice versa.

In addition, proficient data dissemination can be obtained using cooperative transmission. This new communication principle can overcome connectivity problems in sparse settings or heavily partitioned topologies. With cooperative transmission, a group of nodes can combine its emission power and achieve a higher cumulative emission power. Cooperatively transmitting nodes emit identical symbols synchronously, and thus, by superimposing the emitted waves on the physical medium the destination will receive the sum of waves, resulting in a higher total power. That is how the nodes can reach destinations that are very far away [47].

In [47], a continuously changing environment is considered and four types of communication principles are analyzed: (i) traditional multi-hop communication (flooding); (ii) wave propagation cooperative transmission; (iii) accumulating cooperative transmission; and (iv) ideal hybrid multi-hop cooperative transmission.

In sparse settings, the multi-hop transmission can fail when two nodes that need to transfer the message to the sink are separated more than their transmission range. The second approach is based on the principle of Opportunistic Large Arrays (OLA), introduced in [48]. OLA is a form of cooperative diversity in which a large group of simple, inexpensive relays operate without any mutual coordination, but naturally transmit at approximately the same time in response to energy received from the single source (or another OLA). In the wave propagation cooperative transmission approach, each node will repeat a received message once. It will do this together with all other nodes who have received the same message at the same time. Thus, the message will propagate through the network like a wave front. The transmission can terminate before it reaches the sink, if after the first steps, only few nodes are included in the cooperative transmission, and the distance to the next nodes is large, thus unachievable.

Accumulative cooperative transmission is a modification of the previous approach, which goes beyond the previous problem. Here, the nodes that have received a message will transmit the message several times (number of retransmissions will be set as a system parameter), thus the probability that the message reaches the sink will be also increased.

The fourth approach proposes alternating the communication between multi-hop and accumulative cooperative transmission. After the cooperative transmission, all "new" nodes will try to acquire further partners using multi-hop and, after this, a new accumulated cooperative transmission will take place with the now larger group.

The overall coverage can be improved with this mechanism which allows lost nodes or clusters to establish a connection in a new way, not possible without cooperative transmission. Additionally, for the delivery and relaying of packets, it is not necessary to keep track of connections and paths. Nodes will simply repeat a message several times after reception.

6.5 Conclusion

This chapter reviews the most recent proposals utilizing sensor nodes mobility and data dissemination techniques in WSNs. The mobility of the sensor nodes is an important aspect in the future WSNs as it may improve the monitoring capabilities and coverage, enhance communication, and enable new applications. The mobility of network components can deliver results that are greatly beneficial to the network performances.

Research on mobility in WSNs evokes a lot of open issues. They comprise designing novel algorithms and protocols, providing adequate performance metrics and performance analysis and designing new and appropriate mobility models. Additional complexity is introduced by allowing different mobility models to have different influence on the performance metrics. Therefore, the research community is actively working toward these challenges.

References

1. Coskun V (2008) Relocating sensor nodes to maximize cumulative connected coverage in wireless sensor networks. Sensors 8:2792–2817, ISSN 1424-8220
2. Gupta P, Kumar PR (1998) Critical power for asymptotic connectivity in wireless networks. In: Stochastic analysis, control, optimization and applications., pp 547–566
3. Wang X, Xing G, Zhang Y, Lu C, Pless R, Gill C (2003) Integrated coverage and connectivity configuration in wireless sensor networks. ACM SenSys, New York
4. Lopez TS, Kim D (2008) Wireless sensor networks and RFID integration for context aware services. Technical Report, Auto-ID Labs White Paper, White Paper series
5. Efthymiou C, Nikoletseas S, Rolim J (2006) Energy balanced data propagation in wireless sensor networks. Wirel Netw (WINET) J 12(6):691–707
6. Wang W, Srinivasan V, Chu KC (2005) Using mobile relays to prolong the lifetime of wireless sensor networks. In: MobiCom '05, Proceedings of the 11th annual international conference on Mobile computing and networking. ACM, Cologne, Germany
7. Jim Butler (May 2003) Robotics and microelectronics: mobile robots as gateways into wireless sensor networks. In: Proceeding of Technology@Intel Magazine
8. Adireddi S, Tong L, Zhao Q (2003) Sensor networks with mobile agents, vol 22. In: Proceeding of IEEE MILCOM, Boston, MA, pp 688–693
9. Chatzigiannakis I, Kinalis A, Nikoletseas S (2008) Efficient data propagation strategies in wireless sensor networks using a single mobile sink. Comput Commun (COMCOM) J 31(5):896–914, Elsevier
10. Luo J, Hubaux JP (2005) Joint mobility and routing for lifetime elongation in wireless sensor networks. In: Proceeding of INFOCOM, 24th annual joint conference of the IEEE computer and communications societies, Miami, FL
11. Chandrakasan A, Sabhrwal A, Aazhang B (2003) Using predictable observer mobility for power efficient design of sensor networks. In: The second International Workshop on Information Processing in Sensor Networks (IPSN)
12. Kansal A, Somasundara A, Jea D, Srivastava M, Estrin D (2004) Intelligent fluid infrastructure for embeded neworks. In: Proceedings of the 2nd international conference on Mobile systems, applications, and services (MobiSYS)
13. Shah R, Roy S, Jain S, Brunette W (2003) Data mules: modeling a three-tier architecture for sparse sensor networks. In: Proceeding of the IEEE workshop on sensor network protocol and applications (SNPA)
14. Zhao W, Ammar M (2003) Message ferryng: proactive routing in highly-partitioned wireless ad hoc networks. In: Proceedings of the 9th IEEE workshop on future trends of distributed computing systems (FTDCS'03)
15. Zhao W, Ammar M, Zegura E (2004) A messare ferrying approach for data delivery in sparse mobile ad hoc networks. In: Proceeding of the 5th ACM international symposium on Mobile ad hoc networking and computing (MobiHoc). Roppongi Hills, Tokyo, Japan, ACM Press, pp 187–198
16. Bartolini N, Calamoneri T, Fusco EG, Massini A, Silvestri S (2008) Snap and spread: a self-deployment algorithm for mobile sensor networks. In: Proceedings of the 4th IEEE/ACM international conference on Distributed Computing in Sensor Systems (DCOSS)., pp 451–456
17. Li X, Nayak A, Simplot-Ryl D, Stojmenovic I (2010) Sensor placement in sensor and actuator networks. In: Nayak A, Stojmenovic I (eds) Wireless sensor and actuator networks: algorithms and protocols for scalable coordination and data communication. Wiley, ISBN: 978-0-470-17082-3
18. Howard A, Mataric MJ, Sukhatme GS (2002) Mobile sensor network deployment using potential fields: a distributed, scalable solution to the area coverage problem. In: Proceedings of the 6th international symposium on distributed autonomous robotics systems (DARS), Fukuoka, Japan

19. Poduri S, Sukhatme GS (2004) Constrained coverage in mobile sensor networks. In: Proceedings of IEEE International Conference on Robotics and Automation (ICRA'04), New Orleans, LA, pp 40–50
20. Zou Y, Chakrabarty K (2003) Sensor deployment and target localization based on virtual forces. In: IEEE INFOCOM. Duke University, Durham, NC
21. Heo N, Varshney PK (2005) Energy-efficient deployment of intelligent mobile sensor networks. Proc IEEE TSMCA 35(1):78–92
22. Wang G, Cao G, La Porta T (2004) Movement-assisted sensor deployment. In: Proceedings of the 23rd joint conference of the IEEE computer and communications societies (INFOCOM), vol 4. The Pennsylvania State University, University Park, PA, pp 2469–2479
23. Wang G, Cao G, La Porta T (Nov 2003) A bidding protocol for deploying mobile sensors. In: 11th IEEE International Conference on Network Protocols (ICNP)
24. Wang G, Cao G, La Porta T (2004) Proxy-based sensor deployment for mobile sensor networks. In: Proceedings of the 1st IEEE international conference on Mobile Ad-hoc and Sensor Systems (MASS), pp 493–502
25. Li X, Santoro N (2006) ZONER: a ZONE-based sensor relocation protocol for mobile sensor networks. In: Proceedings of the 6th IEEE international workshop on Wireless Local Networks (WLN). Carleton University, Ottawa, pp 923–930
26. Li X, Santoro N, Stojmenovic I (2007) Mesh-based sensor relocation for coverage maintenance in mobile sensor networks In: Proceedings of the 4th international conference on Ubiquitous Intelligence and Computing (UIC), pp 696–708
27. Yang S, Minglu Li, Wu J (2007) Scan-based movement-assisted sensor deployment methods in wireless sensor networks. IEEE Trans Parallel Distrib Syst 18(8):1108–1121
28. Wang G, Cao G, La Porta T, Zhang W (2005) Sensor relocation in mobile sensor networks. In: Proceedings of the 24rd annual joint conference of the IEEE computer and communications societies (INFOCOM). Pennsylvania State Universtiy, University Park, PA, pp 2302–2312
29. Mousavi H, Nayyeri A, Yazdani N, Lucas C (2006) Energy conserving movement-assisted deployment of ad hoc sensor networks. IEEE Commun Lett 10(4):269–271
30. Li X, Frey H, Santoro N, Stojmenovic I (2008) Localized sensor self-deployment with coverage guarantee. ACM SIGMOBILE Mobile Comput Commun Rev 12((2):50–52, MobiHoc 2007 Poster
31. Butler, Rus D (2003) Event-based motion control for mobile-sensor networks. IEEE Pervasive Comput 2(4):34–42
32. Bisnik N, Abouzeid A, Isler V (2006) Stochastic event capture using mobile sensors subject to a quality metric. Mobicom in Proc. of ACM MobiCom, Los Angeles, CA, USA, pp 98–109
33. Garetto M, Gribaudo M, Chiasserini C-F, Leonardi E (2007) A Distributed Sensor Relocation Scheme for Environmental Control. In: Proceedings of the 4th IEEE international conference on Mobile Ad-hoc and Sensor Systems (MASS). University of Torino, Torino
34. Gavrilovska L, Prasad L (2006) Ad-hoc networking towards seamless communication. Springer, ISBN: 978-1-4020-5065-7
35. Intanagonwiwat C, Govindan R, Estrin D, Heidemann J, Silva F (Feb 2003) Directed diffusion for wireless sensor networking. Proc IEEE/ACM Trans Netw 11(1):2–16
36. Heinzelman W, Chandrakasan A, Balakrishnan H (2000) Energy-efficient communication protocol for wireless microsensor networks. In: proceedings of the 33rd international conference on system science (HICSS'00), Hawaii, Jan
37. Tien Nguyen L, Defago X, Beuran R, Shinoda Y (Oct 2008) An energy efficient routing scheme for mobile wireless sensor networks. In: Proceedings of IEEE International Symposium on Wireless Communication Systems, ISWCS '08., pp 568–572
38. Luo H, et al (2003) TTDD: two-tier data dissemination in large-scale wireless sensor networks. ACM J Mobile Net Appl, Special Issue on ACM MOBICOM '03. A Two Tier Data Dissemination Model for Large scale Wireless Sensor Networks. MOBICOM'02, September 23–26, 2002, Atlanta, Georgia, USA, pp 148–159

39. Ratnasamy S, et al (Sept 2002) GHT: a geographic hash table for data-centric storage in sensornets. In: Proceedings of 1st ACM international workshop on wireless sensor network and applications WSNA 2002, Atlanta, Georgia, USA, September 28, 2002. ACM 2002, ISBN 1-58113-589-0, pp 78–87
40. Hamida EB, Chelius G, "A Line-Based Data Dissemination Protocol for Wireless Sensor Networks with Mobile Sink," in *Proceedings of IEEE ICC '08*, Beijing, China, May 2008
41. Lu J-L, Valois F (2007) On the data dissemination in WSNs. 3rd International Conference on Wireless and Mobile Computing, Networking and Communications, New York, Oct
42. Hamida EB, Chelius G (2008) Strategies for data dissemination to mobile sinks in wireless sensor networks. In: Proceedings of IEEE wireless communications 15(6), Dec, pp 31–37
43. Wang Y, Wu H (2006) DFT-MSN: the delay/fault-tolerant mobile sensor network for pervasive information gathering. INFOCOM 2006, Barcelona, Spain
44. Wang Y, Yu H (2007) Delay/Fault-Tolerant Mobile Sensor Network (DFT-MSN): a new paradigm for pervasive information gathering. IEEE Trans Mob Comput 6(9):1021–1034
45. Juang P, Oki H, Wang Y, Martonosi M, Peh L, Rubenstein D (2002) Energy efficient computing for wildlife tracking: design tradeoffs and early experiences with zebranet. In: 10th ASPLOS. ACM Press, San Jose, CA
46. Kinalis A, Nikoletseas S (2008) Adaptive redundancy for data propagation exploiting dynamic sensor mobility. In: MSWiM 2008., pp 149–156
47. Krohn A et al (2006) Increasing connectivity in wireless sensor network using cooperative transmission. INSS 2006, Chicago, IL, 31 May–2 June
48. Scaglione A, Hong YW (Aug 2003) Opportunistic large arrays: cooperative transmission in wireless multihop adhoc networks to reach far distances. IEEE Trans Signal Process 51(8): 2082–92

Chapter 7
Modeling for Geo-Routing in Multi-Hop Wireless Networks

Adnan Khan, Costas Constantinou, and Ivan Stojmenovic

Existing routing and broadcasting protocols for ad hoc networks assume an ideal physical layer. In reality, an accurate representation of physical layer is required for analysis and simulation of multi-hop networking in sensor and ad hoc networks. This work describes a physically realistic model for the log-normal correlated shadow fading loss, and investigates the importance of the shadowing correlation length on designing protocols for ad hoc and sensor networks. Nodes that are geographically proximate often experience similar environmental shadowing effects and can have correlated fading. This work also describes an evaluation procedure, which produces statistically meaningful results over a large number of realizations of multi-hop networks. This procedure takes into account the overall path loss (shadow fading and median path loss) based for antennas working at 2.4 GHz with heights ranging from 0.5 to 1.8 m. Finally, we analyze and compare the performance of the localized position-based greedy algorithm used for Unit Disk Graph (UDG) and the probabilistic position-based greedy algorithm on the proposed model for different values of standard deviation (σ) of shadow fading to show the importance of both the shadow fading and correlation length while designing the protocols for ad hoc networks.

7.1 Introduction

The operation, performance, and limitations of multi-hop wireless networks cannot be understood through theory and simulations based on ever-improving models of point-to-point radio propagation channels. This assertion is intuitively self-evident, as the operation of a spatially extended wireless peer-to-peer network is contingent on the spatio-temporally correlated (near) error-free operation of multiple wireless point-to-point links that are "embedded" in the same geographical space.

It has long been recognized that the degree of realism of the radio propagation environment in multi-hop networks needs to be significantly improved.

A. Khan, C. Constantinou, and I. Stojmenovic (✉)
University of Birmingham, UK
e-mail: stojmenovic@gmail.com

L. Gavrilovska et al. (eds.), *Application and Multidisciplinary Aspects of Wireless Sensor Networks*, Computer Communications and Networks,
DOI 10.1007/978-1-84996-510-1_7, © Springer-Verlag London Limited 2011

Early attempts simply aimed to improve the realism with which a point-to-point communication model has been implemented [1–3].

Numerous publications, both in the radio engineering [4–6] and wireless networking communities [13, 15], have identified the need to move away from conceptually simple mathematical models of the wireless channel. Starting from the simplest of such models, it is widely accepted that the unit disk graph (UDG) model, much favored by theoreticians in the computer science community, results in algorithms and predictions that deviate significantly from reality as verified by experiments. Our definition of the UDG model is taken to be

$$p(\text{radio packet reception}) = 1 \ \text{ if } r \leq R$$
$$p(\text{radio packet reception}) = 0 \ \text{ if } r > R$$

where p denotes probability, r is the distance between two radio nodes, and R is a notional coverage radius, possibly specific to each transmitting node, but more frequently identical for all nodes in a wireless peer-to-peer network.

7.2 Review of Basic Radio Wave Propagation Mechanisms and Models

By definition, the received power by an antenna can be expressed as

$$\begin{pmatrix}\text{Received power}\\ \text{in Watts}\end{pmatrix} = \begin{pmatrix}\text{radiated power intensity}\\ \text{at receiver}\\ \text{in Watts/metre}^2\end{pmatrix} \times \begin{pmatrix}\text{effective receiver}\\ \text{antenna aperture area}\\ \text{in metre}^2\end{pmatrix}$$

The principle of reciprocity can be applied to a pair of antennas to show that, the ratio of the gain to the effective aperture area is a universal constant for all antennas of a given frequency, f (in Hz), and wavelength, $\lambda = c / f$ (in m), where c (in ms^{-1}) is the speed of light. This universal constant can be explicitly computed for a simple dipole to give

$$\frac{G}{A_e} = \frac{4\pi}{\lambda^2}$$

The law of conservation of energy between two antennas situated in free-space (vacuum) then takes the following form:

$$P_{\text{rx}} = \left\{ (P_{\text{tx}} G_{\text{tx}}) \left(\frac{1}{4\pi d^2} \right) \right\} \left(\frac{G_{\text{rx}} \lambda^2}{4\pi} \right)$$

where the last term on the right-hand side is the receiving antenna effective aperture, the term in the curly brackets is the radiated power intensity at the receiving antenna; the first term in the curly brackets is the effective, isotropically radiated power (since for an isotropic antenna $G_{tx} = 1$ the effective power is the same as the radiated power), in the direction of the receiving antenna, by the transmitter; and the second term in the curly brackets is the inverse-square law, a manifestation of the principle of conservation of energy, which states that the energy is evenly spread over the surface of a sphere whose center is at the transmitter and the surface contains the receiver at a radial distance d from the transmitter.

It is common practice for engineers to work using logarithmic units (decibels). The conversion of a power ratio from linear units to decibels is

$$r = \frac{P_2(\mathrm{W})}{P_1(\mathrm{W})} \Rightarrow R(\mathrm{dB}) = 10\log_{10} r = 10\log_{10}\frac{P_2}{P_1}$$
$$\Rightarrow R = 10\log_{10}(P_2(\mathrm{W})) - 10\log_{10}(P_1(\mathrm{W}))$$

The convention is to take the unit of power out of the logarithm by subtracting and adding $10\log_{10}(1\ \mathrm{W})$and expressing terms such as

$$10\log_{10}(P_2(\mathrm{W})) - 10\log_{10}(1(\mathrm{W})) = 10\log_{10}\left(\frac{P_2}{1}\right)(\mathrm{dBW}) = 10\log_{10}(P_2)(\mathrm{dBW}).$$

where now we introduce the "unit" of dBW, or decibel relative to 1 W. Decibels relative to 10^{-3} W = 1 mW are expressed as dBm, or decibels relative to 1 mW.

Taking logarithms on both sides of the earlier equation for conservation of energy, yields

$$P_{rx}(\mathrm{dBW}) = P_{tx}(\mathrm{dBW}) + G_{tx}(\mathrm{dBi}) + G_{rx}(\mathrm{dBi}) - 10\log_{10}\left(\frac{4\pi d}{\lambda}\right)^2$$

where we have normalized the logarithmic antenna gains relative to the gain of an isotropic antenna, which is by definition equal to unity, and introduced the unit of dBi, or decibel relative to an isotropic antenna. The last term on the right-hand side is identified to be the free-space path loss, $L_0(\mathrm{dB}) = 20\log_{10}(4\pi d/\lambda)$.

Empirical-statistical propagation models are a generalization of the above expression and typically take the following form [7]:

$$P_{rx}(\mathrm{dBW}) = P_{tx}(\mathrm{dBW}) + G_{tx}(\mathrm{dBi}) + G_{rx}(\mathrm{dBi}) - L(\mathrm{dB})$$

where $P_{tx}(\mathrm{dBW})$ is the transmitted power in decibels relative to 1 W, $P_{rx}(\mathrm{dBW})$ is the corresponding received power, $G_{\{tx,rx\}}(\mathrm{dBi})$ is the transmit/receive antenna gain in the direction of the receiver/transmitter measured in decibels relative to an idealized reference isotropic antenna, and L(dB) is the path loss between the two antennas in decibels. As we shall see shortly, for nearly all propagation environments of any practical interest we have highly incomplete knowledge of L(dB), which we separate into an empirical, deterministic component and one or more stochastic components.

In the case of free-space path loss, we have been considering above, the free-space loss is deterministic and can be expressed simply as

$$L_0(\text{dB}) = 20\log_{10}\left(\frac{4\pi d}{\lambda}\right) = 20\log_{10}\left(\frac{4\pi df}{c}\right) = 20\log_{10}\left(\frac{4\pi}{c}\right) - 20\log_{10} f - 20\log_{10} d$$

Once again, it is conventional for radio propagation engineers to use MHz and km instead of Hz and m in describing frequencies and distances respectively, so the convention is to label the relative quantities with a subscript denoting the use of non-SI units (rather than adopt units such as a dBMHz). Doing so involves changing the conventional units of measuring the speed of light in vacuum from $c = 3\times10^8\ \text{ms}^{-1}$ to $c = 0.3\ \text{km}\cdot\text{MHz}$, which renders the above expression in its commonly used form in engineering,

$$L(\text{dB}) \equiv L_0(\text{dB}) = 32.4 + 20\log_{10} f_{\text{MHz}} + 20\log_{10} d_{km}$$

where f_{MHz} is the radio wave frequency in MegaHertz and d_{km} is the antenna separation in kilometers.

A second example of a simple, deterministic path-loss example occurs when the two radio nodes are situated over a flat ground. For sufficiently long antenna separations, geometrical optics approximations apply and the method of images can be used together with the fact that at near-grazing incidence the ground reflection coefficient is asymptotically approaching the value −1, to give

$$L(\text{dB}) = 40\log_{10} d - 20\log_{10} h_{\text{tx}} - 20\log_{10} h_{\text{rx}} \quad \text{valid for} \quad d > 4h_{\text{tx}}h_{\text{rx}}/\lambda$$

where $h_{\{\text{tx},\text{rx}\}}$ is the transmitter/receiver antenna height over ground measured in the same units as the antenna separation d and λ is the radio carrier wavelength also measured in the same units.

For more general radio propagation environments where multipath propagation can occur and obstructions can give rise to spatially extended geographical areas of reduced average signal strength (these correspond to geometrical optics shadow regions), we resort to nondeterministic expressions for the point-to-point path-loss equation. Such expressions have been empirically derived and verified through numerous measurement campaigns for land mobile radio systems and in their majority cannot be extrapolated to low antenna heights applicable to mobile ad hoc networking research [4, 8].

Such path-loss models typically take the form,

$$L(\text{dB}) = \{A + 20\gamma\log_{10} d - \alpha h_{\text{tx}} - \beta h_{\text{rx}}\} + \{x_\sigma\} + \{y\}$$

where the parameters A, α, β, and the path-loss exponent γ are empirically found to be frequency-dependent and change depending on the environment (e.g., urban, rural, open), whether there exists a line-of-sight (LOS) between the transmitter and receiver and also may only be valid for specific ranges of the separation d. The determination of these parameters is made through least-squares fitting of the above expression on data obtained from extensive measurement campaigns in "typical" propagation environments. Such an expression is only valid for the class of environments in which the measurements were performed and in the case of [4] which we shall use in this work for low antenna heights as terms depending on the logarithm of the antenna heights are more commonplace when at least one of these heights above the ground is significant. Unlike cellular radio systems, sensor networks have low-lying antennas and only few empirical propagation models apply to them.

Finally, the last two terms, in curly brackets, x_σ and y are zero mean random variables taken from a statistical distribution that describes shadow fading and multipath fading, respectively. The first term in the curly brackets is commonly referred to as the area mean path loss, the second term is referred to as the local mean, shadowing or slow fading, and the third term is referred to as the fast or multipath fading. The local mean, x_σ, arises from the fact that a statistically significant number of measurement locations occurs when large (on the linear scale of a wavelength λ) obstacles exist between the transmitter and the receiver, severely limiting the amount of radiation intensity at the receiver. At each and every location, the variable x_σ takes definite values that are dependent on the specific buildings and ground undulations of the area surrounding both the transmitter and receiver, but in the absence of accurate topography information, it is treated as a stochastic variable. This stochastic variable x_σ is empirically, typically found to obey a log-normal distribution,

$$p(x_\sigma) = \frac{1}{\sigma_{\mathrm{dB}}\sqrt{2\pi}} \exp\left\{-x_\sigma^2 / 2\sigma_{\mathrm{dB}\,\mathrm{dB}}^{2\;\;2}\right\}$$

where σ_{dB} is the shadowing standard deviation measured in decibels. The reasons for which x_σ obeys a normal (Gaussian) distribution with logarithmic units are not well-understood, although plausibility arguments loosely based on the law of large numbers have been advocated at times.

The multipath fading which arises from the constructive and destructive interference of a large number of waves scattered by different objects in the propagation environment, y, is found to obey various statistical distributions, e.g., Rayleigh, Rician, Nakagami, Weighbull that are strongly environment-dependent (e.g., if there is a dominant scattered wave component, if there is diffuse scattering) and its length scale of variation is of the order of the wavelength λ.

For the purposes of our discussion, it is important to point out that, when the receiver is moving along a trajectory, x_σ and y vary on significantly different spatial scales. The local mean variation is empirically found to be spatially correlated. Taking many measured instances of x_σ as a function of distance along the trajectory allows us to compute the ensemble averaged autocorrelation function $\langle x_\sigma(d) x_\sigma(d+\Delta d)\rangle / \langle x_\sigma^2 \rangle$, which is empirically found to be approximated well by an exponential function $\exp\{-\Delta d/\delta\}$. The e^{-1} correlation distance δ is related to the 50% correlation distance $d_{50\%}$ [6] by $d_{50\%} = \delta/\ln 2$. As is expected, this is strongly environment-dependent (e.g., depends on ground undulations, density, and type of urban/suburban buildings) and normally lies in the range $20\ \mathrm{m} < d_{50\%} < 80\ \mathrm{m}$ in urban environments [4–8].

On the other hand, the multipath fading is known to have a spatial correlation distance of the order of $\lambda/4$ [7], which need not concern us at all, as most radio systems tend to employ fade mitigation diversity techniques that smear out the multipath fading pattern in the immediate vicinity of the receiver.

7.3 Beyond the UDG Model

A simple extension to the UDG is to replace the Heaviside step function behavior of the probability of radio packet reception by a continuous spatial distribution. The straightforward method for achieving this is as follows.

The probability for receiving correctly a packet of length ℓbit for which at most ℓ' bits can be forward error-corrected, can be expressed in terms of the bit-error ratio, p_e, as

$$ppr = \sum_{\lambda=0}^{\ell'} \binom{\ell}{\lambda} p_e^{\lambda} (1-p_e)^{\ell-\lambda}$$

For this work, we assume packets of 128 bits length and use a slightly more complex (7,3) Reed-Muller block coding [10].

The bit-error ratio, p_e, can be determined by the received signal-to-noise ratio and the modulation scheme employed. For example, for quadrature phase shift keying (QPSK), the bit-error ratio is given by [10],

$$p_e = \text{erfc}\left(\sqrt{\frac{E_b}{N_0}}\right) - \frac{1}{4}\text{erfc}^2\left(\sqrt{\frac{E_b}{N_0}}\right)$$

where $\text{erfc}(z) = \frac{2}{\sqrt{}} \int_z^{\infty} \exp\{-t^2\} dt$ is the complementary error function, N_o is the receiver (one-sided) noise spectral density and E_b is the received energy per bit.

Inverting the above formula (only possible numerically), yields the signal-to-noise threshold E_b / N_0 corresponding to a given value of p_e, which is used in computing the outage probability integral [11] which uses the path-loss distribution and the probability density function for x_σ. The computation of such outage integrals is beyond the scope of this chapter and the reader is referred to standard textbooks in the literature [12].

As a consequence, the probability of packet reception can be plotted as a function of the separation of the transmitting and receiving nodes, d, by integrating over all possible values of x_σ(weighted by its probability density function in the integrand) to yield a symmetric coverage model as shown in Fig. 7.1.

In the trivial case when $\sigma_{\text{dB}} = 0$, there is no shadow fading and the probability of packet reception can be computed in closed analytic form. We plot this in Fig. 7.1 for the Konstantinou [4] propagation model, for $h_{\text{tx}} = h_{\text{rx}} = 1.5$ m, $f = 2{,}100$ MHz, $G_{\text{tx}} = G_{\text{rx}} = 2.16$ dBi and a dense urban environment with the published transition model [4] for LOS to non-LOS path-loss ratio, as shown in Fig. 7.2. The propagation model used to plot Fig. 7.2 is summarized here for ease of reference,

$$L_{LOS}(d) = 4.62 + 20\log_{10} 4\pi/\lambda - 2.24h_t - 4.9h_r + 29.6\log_{10} d$$

$$L_{NLOS}(d) = 20\log_{10} 4\pi/\lambda - 2h_r + 40\log_{10} d$$

$$L(d) = \alpha(d)L_{LOS}(d) + (1 - \alpha(d))L_{NLOS}(d)$$

$$\alpha(d) = \begin{cases} \exp\{-(d - a_f)/b_f\} & \text{if } d \geq a_f \\ 1 & \text{if } d < a_f \end{cases}$$

$$\lambda = c/f = 0.143 \text{ m}; \quad \text{for urban areas}, a_f = 5 \text{ m}; \quad b_f = 35 \text{ m}$$

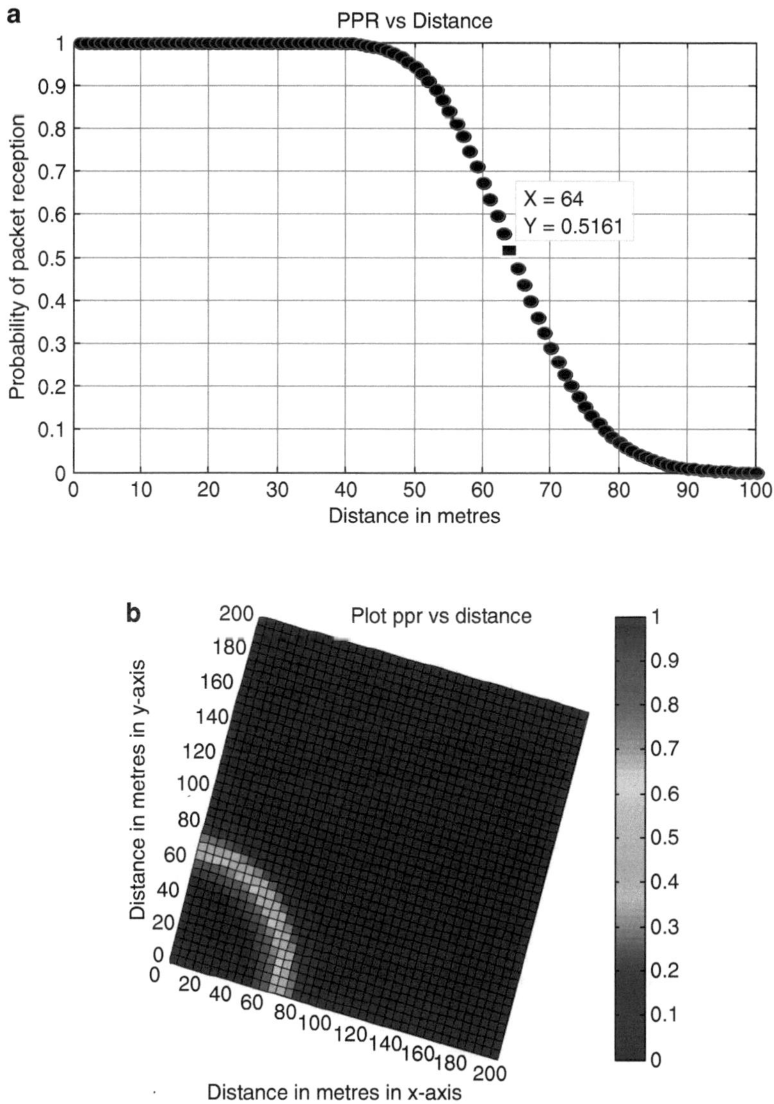

Fig. 7.1 A plot of the probability of packet reception against range in meters from a transmitting node for the model of [4]

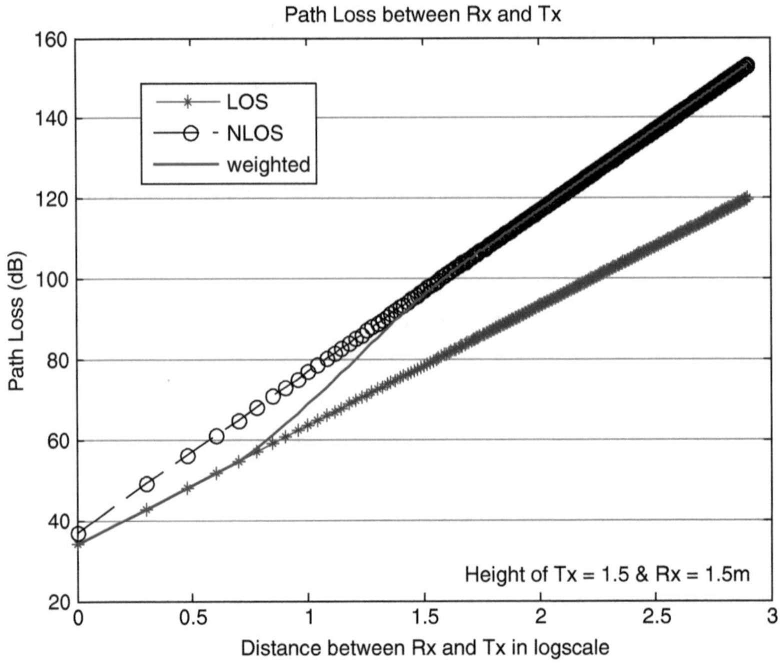

Fig. 7.2 The weighted average path-loss model that combines LOS and NLOS propagation using a smooth transition model as proposed in [4]

We fix the transmitter power to $P_{tx} = -27$ dBWand the receiver sensitivity to -131dBW and its noise floor to $N = -112$ dBW, such that $ppr = 0.5$occurs at $d = 65$ m. These receiver and transmitter characteristics correspond approximately to an IRIS mote specification.

When $\sigma_{dB} > 0$, the packet reception probability at each value of d becomes itself a stochastic variable with a given spread of values around those depicted in Fig. 7.1. The $\sigma_{dB} = 0$ model simply extends the UDG by introducing concentric probability of packet reception contours as shown in Fig. 7.1b, but the $\sigma_{dB} > 0$model cannot account for the fact that the spatial random field, $x_{\sigma}(r)$, possesses a characteristic spatial correlation length, $d_{50\%}$ in a meaningful way, since cumulative outage probabilities will always result in a circularly symmetric model.

7.4 Spatially Correlated Shadow Fading

Modeling the spatially correlated nature of shadow fading can be achieved through a variety of models of varying sophistication [6 and references therein]. We choose a conceptually simple-to-understand method that is not particularly computationally efficient, but is capable of producing correlated shadow fading at points laid out on any arbitrary geographical grid.

It has been widely reported in the literature [5] that the spatial correlation of the shadow fading at two nearly locations $\boldsymbol{r}_i$ and $\boldsymbol{r}_j$ is known to be of the form,

$$\rho_{ij} = \frac{\mathrm{cov}(x_i x_j)}{\sigma^2} = \frac{\langle x_i x_j \rangle}{\sigma^2} = \exp\left(-\frac{|r_i - r_j|}{d_{50\%}} \ln 2 \right) = \exp\left(-\frac{\Delta d_{ij}}{d_{50\%}} \ln 2 \right)$$

where $x_i = x_\sigma(r_i)$, we have taken advantage that the mean value of the shadow fading is by definition zero and we are using the 50% correlation distance, $d_{50\%}$, rather than the e^{-1}correlation distance δ which would result in the omission of the ln 2 factor as we saw earlier.

If we consider a grid consisting of n receiver points arbitrary positioned in space around a transmitter, generating a vector, $\tilde{\boldsymbol{x}}$, of *n i.i.d.* Gaussian random variables to produce an instance of the shadow fading at all the receivers would fail to capture the fact that elements of $\tilde{\boldsymbol{x}}$ that are spatially separated by distances smaller than $d_{50\%}$ are correlated and thus their values should not be *independent* but *correlated* when ensemble averaged. The desirable spatial correlation properties can be captured using a mathematical transformation that converts a vector $\tilde{\boldsymbol{x}}$, of *n i.i.d.* Gaussian random variables with a unit standard deviation to a suitably correlated vector of Gaussian variables with the correct standard deviation σ_{dB}.

We can use the two-point correlation equation given above to construct the $n \times n$ covariance matrix, **C**, such that $c_{ij} = \sigma_{dB}^2 \rho_{ij}$. The Cholesky decomposition [9] of $\boldsymbol{C} = \boldsymbol{LU}$, where $\boldsymbol{C} = \boldsymbol{LU}$, yields the transformation matrix **L** that can be used to transform $\tilde{x}$ to **X**,

$$x = L\tilde{x}$$

The desirable correlation properties the $n \times 1$vector of Gaussian deviates, **X**, can be shown by considering

$$\langle xx^T \rangle = \langle L\tilde{x}\tilde{x}^T L^T \rangle = L \langle \tilde{x}\tilde{x}^T \rangle L^T$$

and since by definition $\tilde{x}$ is an i.i.d. Gaussian process of unit variance and zero mean,

$$\langle \tilde{x}\tilde{x}^T \rangle = I$$

where **I** is the unit $n \times n$ matrix, we have,

$$\langle \boldsymbol{xx}^T \rangle = \boldsymbol{LIL}^T = \boldsymbol{LL}^T = \boldsymbol{C}$$

which shows that the vector **X** has the desired covariance structure. Note that the Cholesky decomposition into the form used above is possible by virtue of the fact that all the matrix elements of **C** are real and positive. An example grid and the corresponding correlated shadow fading field are shown in Fig. 7.3.

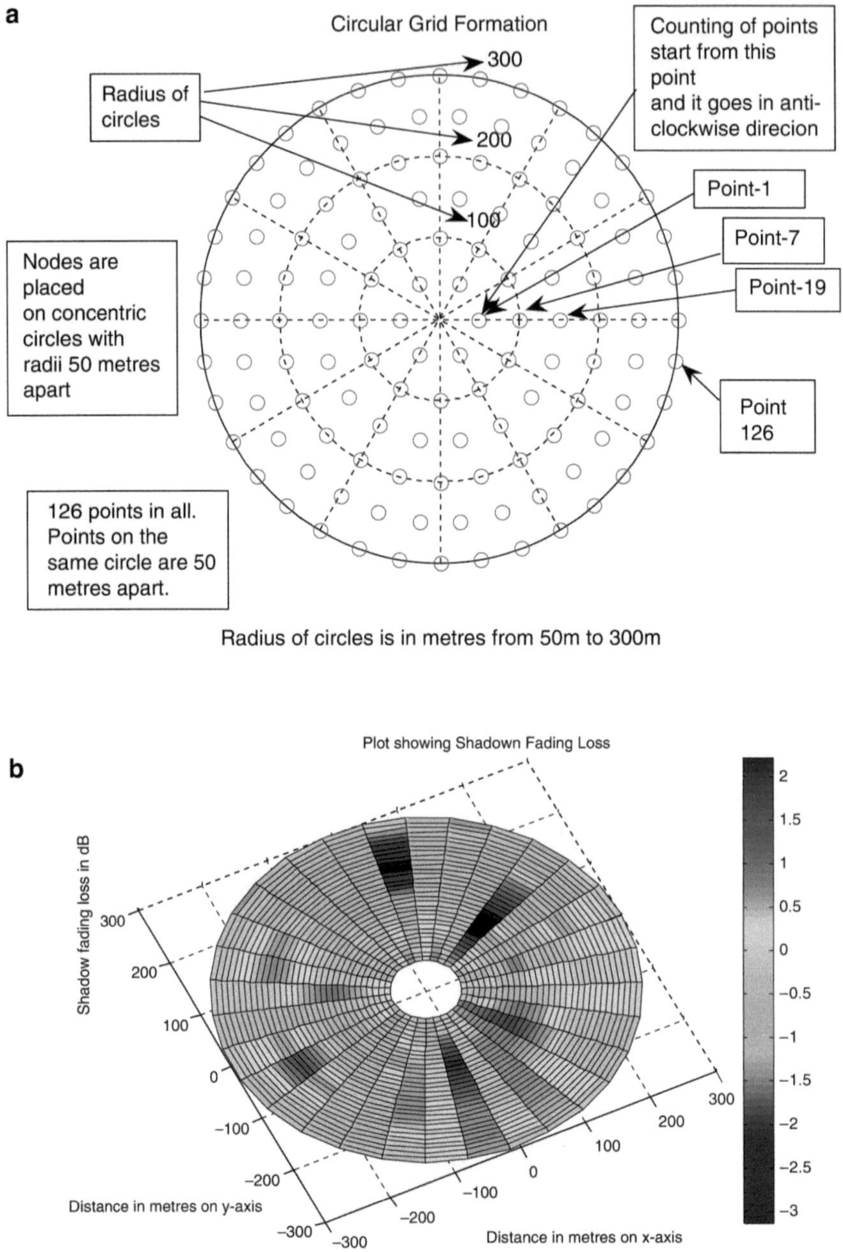

Fig. 7.3 (**a**) The grid over which the correlated shadow fading is computed and (**b**) the spatially correlated shadow fading field, $x_\sigma(\boldsymbol{r})$ (note that this is under-sampled due to the large average grid spacing and is plotted without any smoothing interpolation) with $\sigma_{\text{dB}} = 2\text{dB}$and $d_{50\%} = 70\text{m}$ (or equivalently $d_{1/e} = 100\text{m}$)

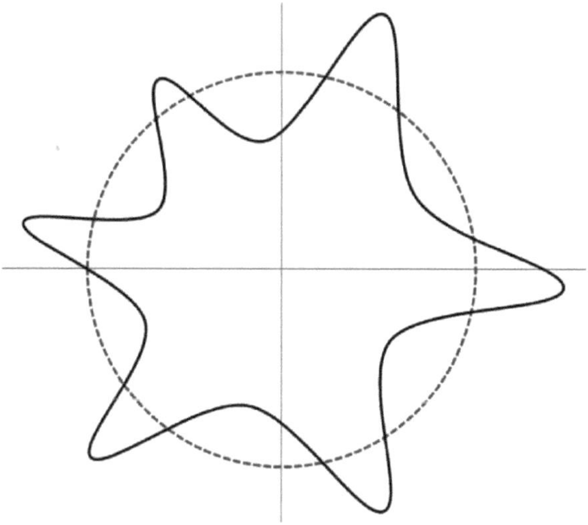

Fig. 7.4 The 100–0% ppr transition contour for the UDG (*dotted line*) is depicted, superimposed on the 50% ppr contour (*solid line*) for a spatially correlated shadow fading environment. *Note that the contours are not accurate or drawn to any scale*

7.5 Packet Reception Probability Contour

For each and every realization of the random field $x_\sigma(\boldsymbol{r})$ such as the one shown in Fig. 7.3b, the packet reception probability will now become an asymmetric contour that resembles an "amoeba" in shape. This is depicted in Fig. 7.4. We proceed to perform simulations to establish under what circumstances the differences between the two packet reception probability models (i.e., the "amoeba" and circularly symmetric models) result in significantly different predictions for the performance of geographic, position-based routing protocols.

7.6 Geographic/Position-Based Routing

In ad hoc networks, the task of finding and maintaining routes is nontrivial since host mobility can result in unpredictable topology changes. In ad hoc networks, "*Localized Routing Algorithms*" have been proposed where nodes do not require the complete network topological information to perform the routing task. More precisely, nodes only require the position of itself and its one-hop neighbors, and position of destination [13].

We assume in this project that each node has a low-power Global Positioning System (GPS) receiver, which provides the position information of the node itself. If GPS is not available, the distance between neighboring nodes can be estimated on the basis of incoming signal strengths. Relative coordinates of neighboring nodes can be obtained by exchanging such information between neighbors [13]. The other assumption, which is made for this project, is that the source node is aware of geographic position of destination. Such algorithms, where the source node is aware of the geographical position of its destination are known as "*Position or Destination*" based routing protocols.

In comparing the two models, the UDG model and the correlated shadow fading model, "*Localized Destination Based*" algorithms have been considered. The reason for comparing the UDG model with the proposed model is that the designers of network layer protocols for ad hoc and sensor networks assume the Unit Disk Graph (UDG) communication model while designing the protocols.

7.6.1 Adopted Unit Disk Graph (UDG) Model

In the UDG model of § 7.1, we choose all nodes to transmit with equal transmission power. Thus all nodes are taken to have a fixed and equal transmission radius *R*. Furthermore, for comparison purposes with the remaining models, we may define an equivalent UDG transmission range, *R*, by either of the two methods given below:

- For adopted propagation path-loss model, we set $\sigma_{dB} = 0\text{dB}$to recover the fully deterministic, circularly symmetric model for the probability of packet reception shown in Fig. 7.1. We then define the *equivalent* UDG radius to be the radial distance, *R*, from the transmitter where $ppr = 0.5$.
- The transmission radius "*R*" can alternatively be chosen on the basis of the node receiver sensitivity, which in the case of IRIS mote is (according to its data sheet [16]) −101 dBm(−131dBW). This transmission radius is calculated only on the basis of mean path loss (i.e., $\sigma_{dB} = 0$ dB) and the received signal strength, without resorting to computation of the probability of packet reception.

The latter method was used in the following simulations, but it is worth pointing out that the difference between the value of the equivalent UDG transmission radius, *R*, given by the above two methods was very small, 63 and 61 m, respectively. Since all nodes are assumed to have the same transmit power and antenna gain and height, *R* is equal for all nodes.

7.6.2 Localized Greedy Algorithm

The localized greedy position-based algorithm as proposed by Finn [14] suggests that the node that is currently holding the packet will forward it to the neighbor that is closest to the destination. This algorithm makes unambiguous and explicit use of the notion of a transmission radius as defined in the idealized UDG model; however, routing fails if no neighbor is closer to the destination than the current node.

Neighbors are the nodes that are defined by the UDG model. The metric used to measure routing optimality in this work is "Hop Count." Hop count can be defined as the number of time a data packet was forwarded on the created route from source to destination node, ignoring potential retransmissions or acknowledgments [15].

An important definition is that of "progress" made while transferring the packet from the one node to another. Let the source node (*S*) be the current node holding the message, "*D*" the destination node, "A" the considered forwarding neighbor. The progress made by forwarding data from $S \rightarrow A$ is

$$\text{progress} = |\boldsymbol{r}_D - \boldsymbol{r}_S| - |\boldsymbol{r}_D - \boldsymbol{r}_A|.$$

In other words, the UDG model will choose the next node among its neighboring (one-hop) nodes that maximizes this progress metric toward its destination [15].

7.6.3 *Localized Probabilistic Progress Algorithm*

This localized algorithm has been proposed in [15] where the probability of packet reception is used to influence its neighbor selection. The *ppr* is calculated as outlined earlier. The probabilistic localized algorithm is a simple extension of the localized greedy algorithm where the neighbor selection criterion metric now becomes $ppr \times \text{progress}$.

Nodes can record the *ppr* through successive measurements of uncorrupted packets from neighbors and either assume that reciprocity holds, or inform their neighbors of the measured *ppr* for each of their set of transmissions over a predefined time window. Alternatively, a node can estimate rather than measure *ppr* from the received signal strength indicator (RSSI) of its radio receiver and employ the intermediate signal-to-noise and bit-error probability models introduced in §7.3, and once again either assume reciprocity or broadcast this information to each of its respective neighbors.

In the simulations that follow we have assumed that the latter case (RSSI + exchange of information) has taken place without any errors, which is an idealized version of what is practically achievable.

7.7 Numerical Simulation Setup

In this section we present briefly how we created a simulation environment for our study. For the simulation we use a 2-dimensional square simulation area and uniformly and randomly distributed *n* nodes in this. The size of the simulation area as well as the number of nodes are varied as described in the next section. The area for the simulation and the number of nodes has been changed to cover a wide range of neighboring node densities. In this study we assumed that all nodes are identical and comply with the specification of IRIS motes [16], as detailed earlier. Another assumption is that the wireless network is in an isolated remote environment with either slow moving or no mobility events. When a sending node tries to measure or estimate channel parameters, it is assumed these channel parameters remain unchanged for time periods longer than the typical packet transmission time as discussed in [17]. The remaining parameters are summarized in Table 7.1.

Before running the UDG model and the proposed correlated shadow fading coverage model, Dijkstra's shortest path scheme was used to test the network connectivity and only connected graphs were used in the simulations [13]. Furthermore, we discard network topologies that cause the simple UDG forwarding model to fail to deliver data packets successfully. The proposed circular grid model shown in Fig. **7.3** which contains an instance of the correlated shadow fading model is produced for every transmitting node in the simulation area separately. Each node is placed in the center of the circular field and the correlated shadow fading can be produced as described in earlier.

Prior to discussing the results of simulations, we illustrate the path choice results of each algorithm outlined in § 7.6, by describing the outcome of the operation of all the algorithms described above in an example 50-node network over a 200×200 m^2 geographical region (Fig. 7.5).

The simulation environment is programmed in Matlab, where the first step is to generate uniformly distributed random node positions with the help of the built-in function "*rand.*" *Rand* command generates values, which lie in the range of "0" and "1" and is initialized with a different seed in each subsequent simulation.

After the generation of the node positions, the transmission radius of UDG, which is found on the basis of receiver sensitivity (−101 dBm), was calculated to be approximately 61 m. All nodes are assigned Cartesian coordinates to represent their locations and the distances to all other nodes are calculated by Euclidean equation.

All nodes that lie within the distance of 61 m, i.e., transmission radius, are one-hop neighbors. All one-hop neighbors will have the "*ppr*" (probability of packet reception) as "1" while outside the transmission radius the "*ppr*" for UDG is taken as "0." In the case of UDG and Dijkstra the one-hop neighbors are exactly the same.

Table 7.1 Chosen simulation parameters

Parameters Used in Simulations	
Height of transmitter antenna	1.5 m
Gain of transmitter antenna	2.16 dBi
Height of receiver antenna	1.5 m
Gain of receiver antenna	2.16 dBi
Power transmitted from transmitter	−27 dBW
Receiver sensitivity	−131 dBW
Transmitter/receiver frequency	2.1 GHz
Packet length	128 bits
BER model	Coherent QPSK
Neighbor selection threshold for probabilistic greedy algorithm	0.5
Degree of Reed–Muller forward error correction coding	3
Path-loss model	[4]

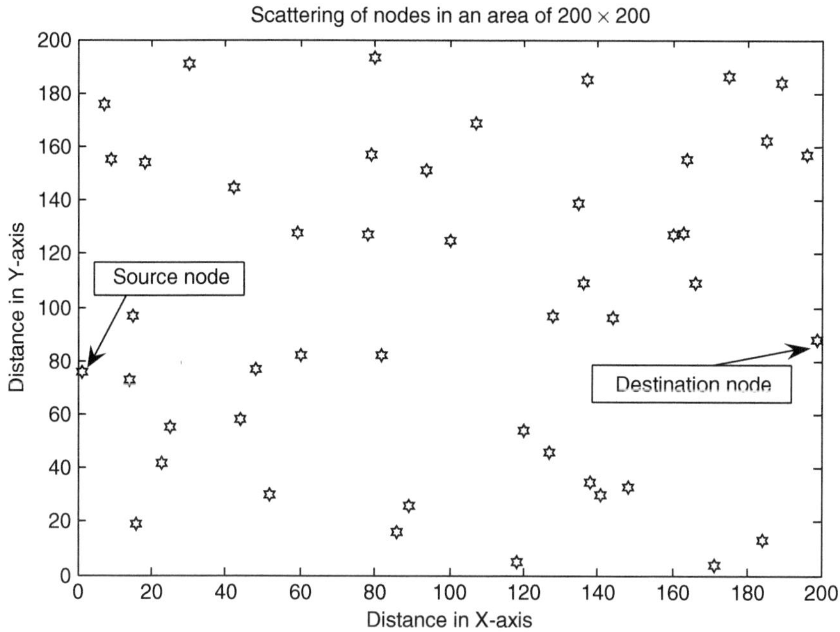

Fig. 7.5 Location of nodes in an area of 200 × 200 m^2

Before running the UDG model, Dijkstra shortest path algorithm was used to test network connectivity, and only connected graphs were used in measurement. If the graph was not connected or if there is no path that exists between source and destination then the model to produce the random nodes generation was run again until a connected graph was found.

Figure 7.6 shows the path selection in the network as given by Dijkstra's algorithm. The blue lines in the plot show the links/paths that can be found between different nodes, while the colored line (black, cyan, red, and green) shows the selected shortest path. In Dijkstra's algorithm every link is normally associated with some cost/weight, here taken to be "1."

Figure 7.7 shows the transmission radius of the optimal path found by Dijkstra's algorithm, which can be seen not to follow the one discovered by the "*Localized Greedy Algorithm.*" This point can be observed by noting the red circle whose center is node 25 and the next forwarding hop is chosen to be 37 instead of node 40 which was the farthest neighboring node in the transmission range of red circle.

After establishing that the network is connected, the UDG model is run on the same topology with the transmission radius "*R*" of 61 m.

Figure 7.8 shows the optimum ideal path that has been obtained by running "*Localized Greedy Algorithm*' which is used for UDG, where the basic functionality has been proved, as the next forwarding neighbor is chosen which is closest to the destination and lies inside the transmission radius "*R*" which is clearly observed from Fig. 7.9.

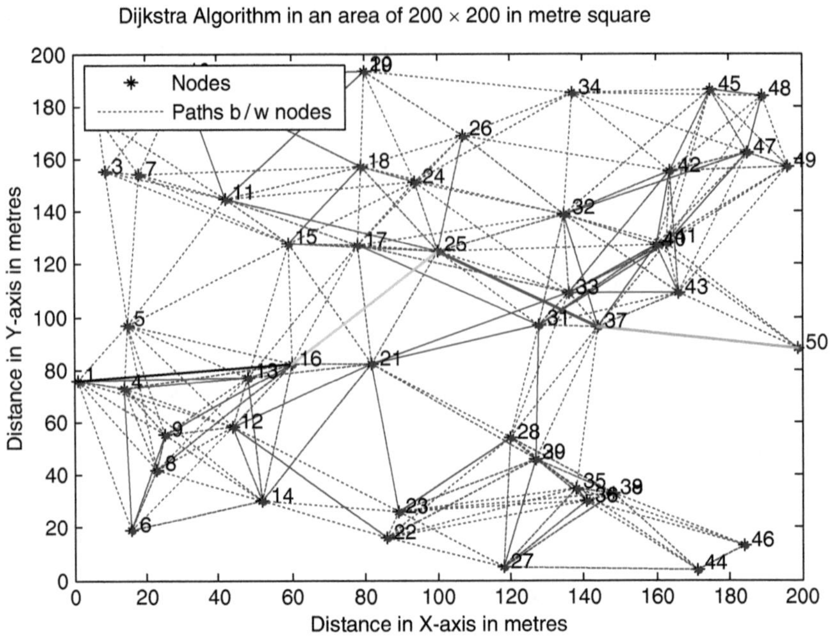

Fig. 7.6 Dijkstra's algorithm showing connectivity in an area of 200 × 200 m² (Hop count = 4)

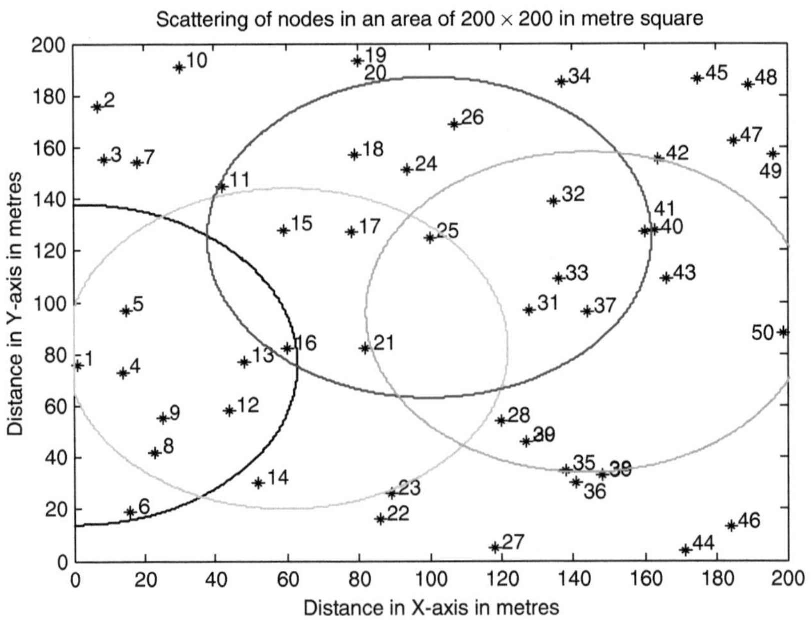

Fig. 7.7 Transmission radius of optimum path of Dijkstra's algorithm in an area of 200 × 200 m²

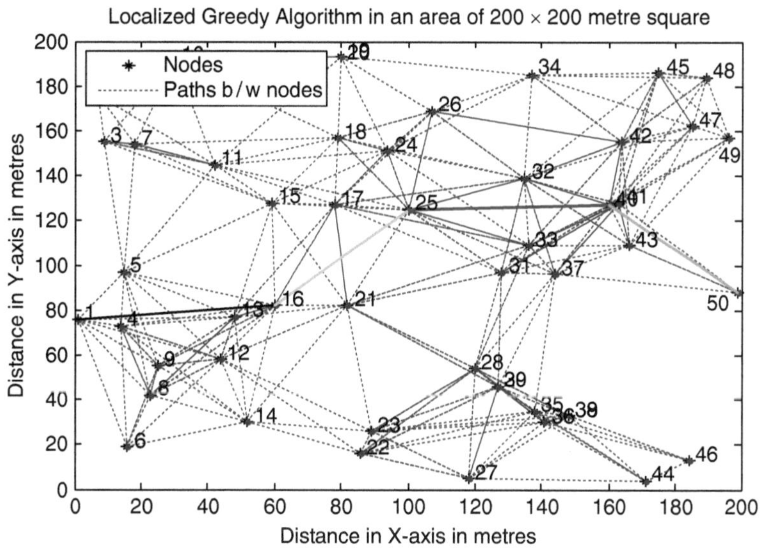

Fig. 7.8 UDG model showing ideal path in an area of 200 × 200 m^2

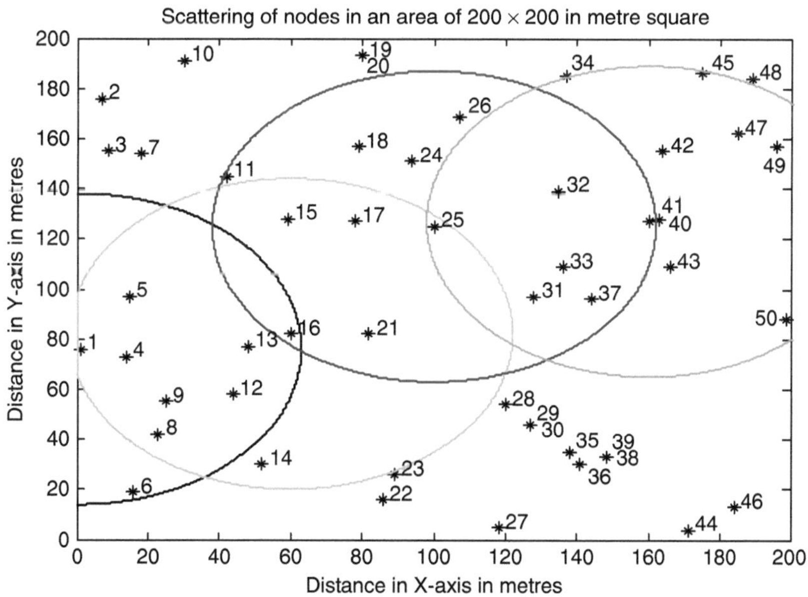

Fig. 7.9 Transmission radius of optimum path of Localized Greedy Algorithm in an area of 200 × 200 m^2

In Fig. 7.9, the source node is "1" and the transmission radius of node "1" is represented with the black circle and the farthest neighbor from node "1" is node "16" which is chosen as the next forwarding hop. Node 16 (transmission radius represented by cyan) then forwards the data to node "'25," as it is closest to the destination (node "50"), and so on.

The two parameters that have been calculated for the Dijkstra and UDG models are the "*Network Density*" and the "*Hop Count.*" Network density is defined as the average number of neighbors per each node that lies within the transmission radius of each node and is found to be 9.72 nodes per UDG area for both UDG and Dijkstra models, since the transmission radius of both is chosen on the basis of receiver sensitivity. Hop count is defined as the number of hops taken to reach from source to destination and is found to be 4 (Figs. 7.6 and 7.8) for both the Dijkstra and UDG models.

7.7.1 *Probabilistic Model When Standard Deviation (σ) of Shadow Fading is Zero*

In this section, the standard deviation of the shadow fading is taken to be zero, which means that the shape of the transmission range will be circular, but still follows the same functionality as defined in pseudo code to find the neighboring nodes and the forwarding hop is chosen on the basis of probabilistic greedy algorithm. With $\sigma = 0$ there is no shadow fading and the probabilistic greedy algorithm works only on the median path loss and the plots of the "*ppr*" against distance is given in Fig. 7.1a.

Figure 7.1a shows the distance that can be included in the transmission range which is around 64 m when the *ppr* is taken to be 0.5 with $\sigma = 0$. The plot shown below shows the path that has been chosen by the probabilistic greedy algorithm for the example of 50 nodes in an area of 200×200 m^2:

The network node density in this case is 10.36 nodes per UDG area and the hop count is 5 (Fig. 7.10).

7.7.2 *Proposed Model When Standard Deviation (σ) of Shadow Fading is Six Decibels*

In this section, the standard deviation of the shadow fading is taken to be 6 dB and follows the same functionality as defined in pseudo code to find the neighboring nodes and the forwarding hop is chosen on the basis of probabilistic greedy algorithm. The path followed from source to destination is given below:

The two metrics that have been calculated on the proposed model with $\sigma = 6$ dB are the "*Network Density*" and the "*Hop Count.*" The network node density in this case is 11.42 nodes per UDG area and the hop count is 4 (Fig. 7.11).

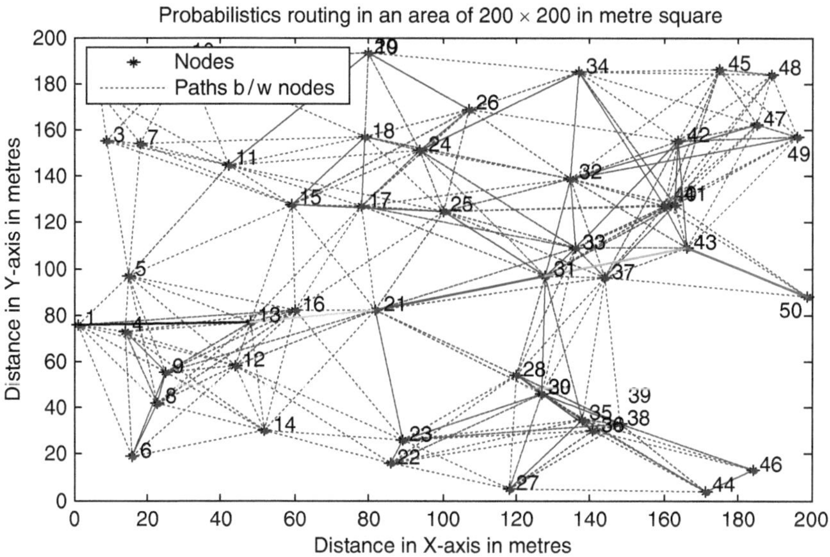

Fig. 7.10 Probabilistic model showing next forwarding hops ($\sigma = 0$) in an area of 200×200 m^2

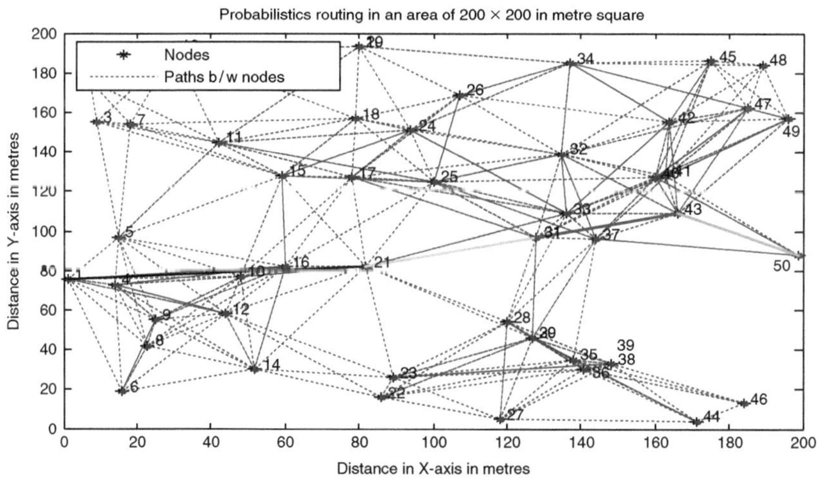

Fig. 7.11 Probabilistic model showing next forwarding hops ($\sigma = 6$ dB) in an area of 200×200 m^2

7.7.3 Proposed Model When Standard Deviation (σ) of Shadow Fading is Eight Decibels

The two metrics that have been calculated on the proposed model with $\sigma = 8$ dB are the "*Network Density*" and the "*Hop Count.*" The network density in this case is 14.90 nodes per UDG area and the hop count is 3 (Fig. 7.12).

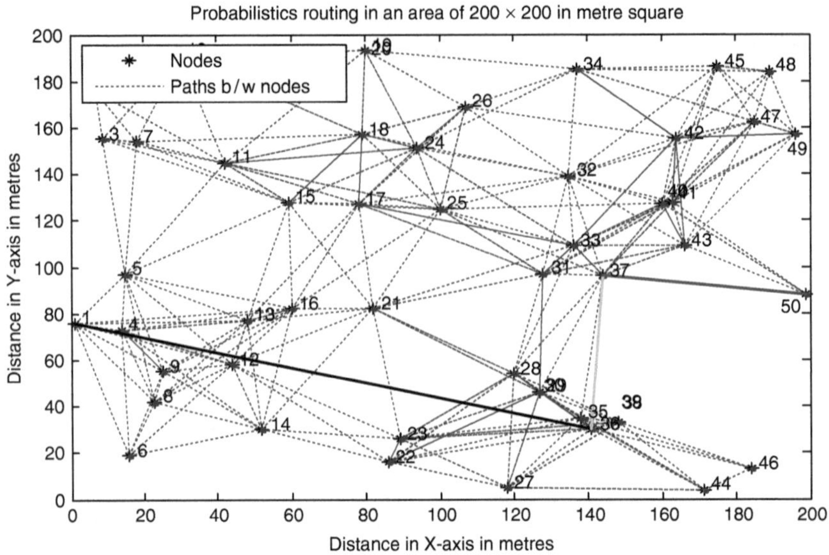

Fig. 7.12 Probabilistic model showing next forwarding hops ($\sigma = 8$ dB) in an area of 200 × 200 m²

7.8 Simulation results

In running the simulations we vary the following parameters:

Parameter 1: Shadowing correlation length to equivalent UDG transmission radius ratio, defined as

$$\psi = \frac{\text{shadow fading 50\% de - correlation distance}}{\text{equivalent UDG radius } (d \text{ for which } ppr = 0.5)} = \frac{d_{50\%}}{R}.$$

This parameter, schematically shown in Fig. 7.13 determines whether the correlated shadow fading *ppr* contour "amoeba" has many protrusions/legs (low value of ψ) or few such features (high value of ψ).

Parameter 2: Average node density, defined as

$$\kappa = \frac{\text{average number of neighbouring nodes for which } ppr \geq \text{threshold} = 0.5}{\text{UDG area}}.$$

The average node density is controlled by varying the number of nodes and the size of the simulation area. The values employed in the simulations were an area of 400 × 400 m² with 100, 125, 135, 150, 200, 250, 275, and 300 nodes and an area of 500 × 500 m² with 250, 275, 300, and 350 nodes. For every such simulation, 12 independent realizations were chosen in order to ensure that statistically meaningful hop count data were collected.

Parameter 3: Shadow fading standard deviation, σ_{dB}. This was chosen to take values of 0 dB (an extreme case of no fading where propagation and *ppr* are

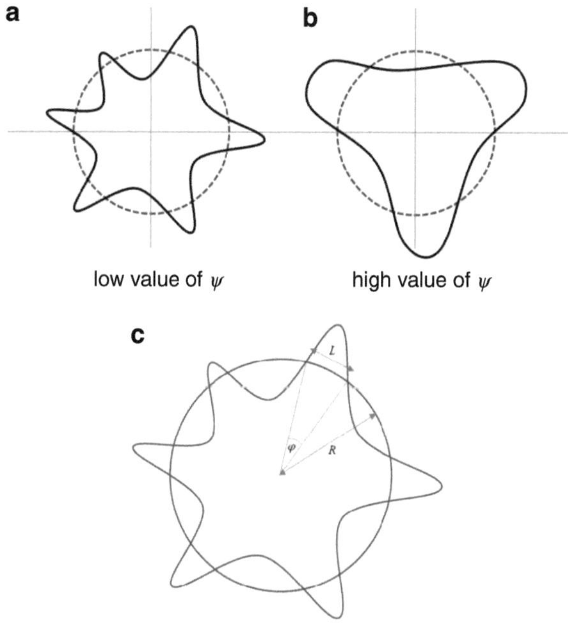

Fig. 7.13 Typical *ppr* contours (*solid lines*) for (**a**) a low value of $\psi < 1$, (**b**) a high value of $\psi > 1$; (**c**) For $\psi \le 1$, the ratio ψ acquires the geometrical explanation of being approximately equal to the angle in radians, as observed at the transmitting node, of the "typical width" of its regions, where $d > d_{50\%} = R = 70$ m for its $ppr = 0.5$ contour, since $\psi = L/R = 2\sin\frac{1}{2}\varphi \approx \varphi$ for $\varphi \le 1^c$(maximum error of 4%)

deterministic functions of distance from the transmitter), 6 and 8 dB, which are the typical reported values in [4].

The simulations are performed as follows: The source node is always chosen to be the node with the minimum x-coordinate value and the destination node is always chosen to be the node with the maximum x-coordinate value throughout the simulations. The average hop count over 12 independent network topology realizations per node density is computed, together with standard error bars to show the spread of values. The results of the simulations are shown in Figs. 7.14–7.16. The greedy algorithm has a hop count close to the optimal (found by Dijkstra's algorithm); local knowledge prevents it from finding always an optimal path, and sometimes it does not find a path at all in a connected network. Furthermore, the performance of the unrealistic $\sigma_{dB} = 0$ dB model is, *on average*, expected to be the same as that of the greedy algorithm on UDG and this is in good agreement with the results of Figs. 7.14–7.16 given that only 12 ensemble realizations are used to compute averages.

It is clear from the simulations that there exist certain combinations of values of ψ, κ and σ_{dB}, where the performance of the probabilistic progress localized routing algorithm with a realistic physical layer is significantly better than the corresponding performance predicted by greedy algorithm applied on both the UDG and the unrealistic $\sigma_{dB} = 0$ dB models.

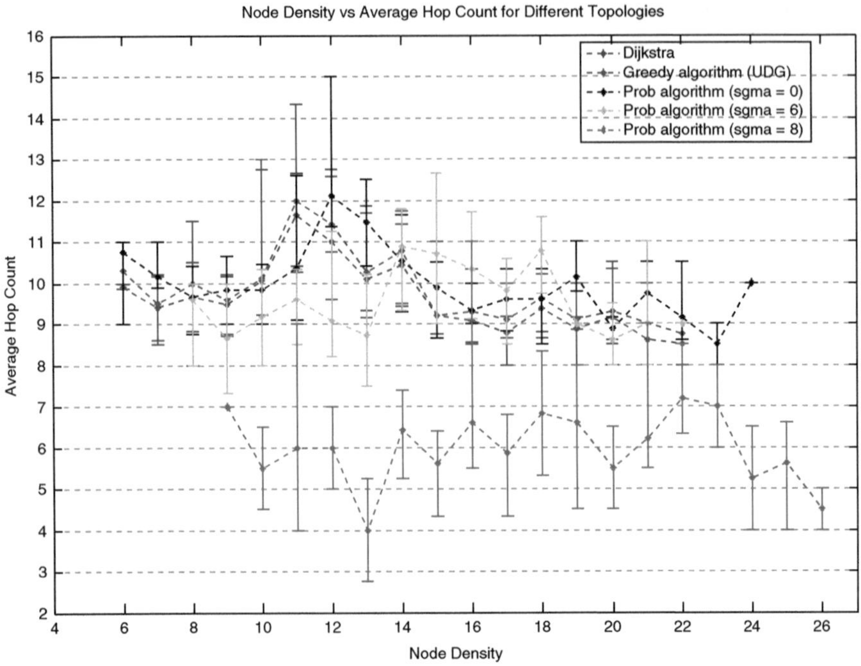

Fig. 7.14 Average hop count against node density for Dijkstra's algorithm, the UDG-based greedy routing and the probabilistic progress localized greedy routing on model employing a realistic physical layer for different values of σ_{dB}, and with a correlation length of 100 m (ψ=1.63)

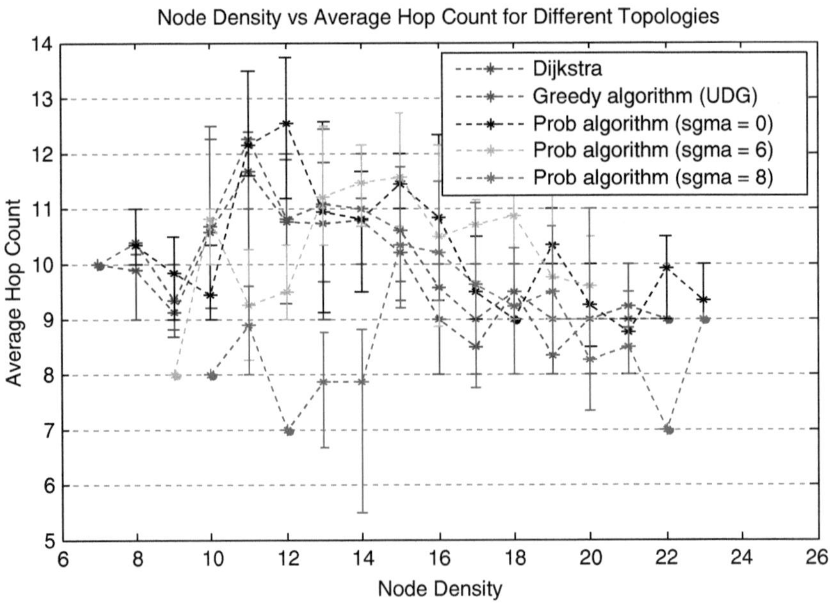

Fig. 7.15 Average hop count against node density for Dijkstra's algorithm, the UDG-based greedy routing and the proposed probabilistic localized greedy routing model employing a realistic physical layer model for different values of σ_{dB}, and with a correlation length of 30 m (ψ = 0.49)

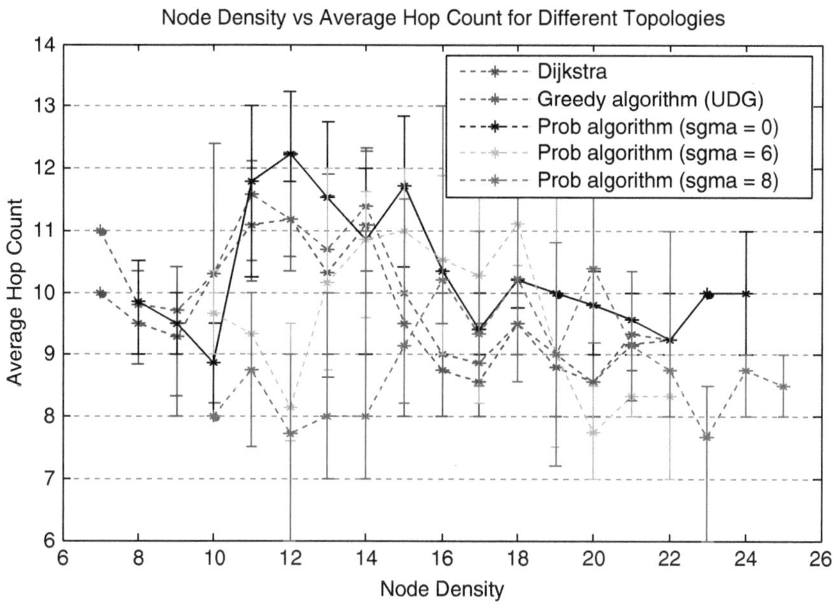

Fig. 7.16 Average hop count against node density for Dijkstra's algorithm, the UDG-based greedy routing and the probabilistic progress localized greedy routing on model employing a realistic physical layer for different values of σ_{dB}, and with a correlation length of 150 m ($\psi = 2.45$)

The performance of the Dijkstra-based algorithm, and greedy algorithm applied on the UDG and the partially physically "realistic" model with no shadow fading $\sigma_{dB} = 0$ dBcan be seen to be comparable throughout.

For values of ψ significantly greater than 1, a narrow range of average local node densities $10 < k < 15$and $\sigma_{dB} \geq 8$dE, we observe a statistically significant reduction in the average hop length of approximately 30%. For values of ψ significantly less than 1, we observe a slightly reduced advantage and only for $\sigma_{dB} \geq 8$dB.

The most important observation though, is that for $\psi = 1$ and $\sigma_{dB} \geq 8$dB, we observe a consistent reduction in the average hop count of approximately 35–40% for all values of average local node density in the simulated range of $9 \leq \kappa \leq 26$. Under these circumstances we can claim with some confidence that the probabilistic progress localized routing algorithm with a realistic propagation model incorporating correlated shadow fading is capable of making next hop decisions that offer substantial performance improvements.

This observation can be plausibly explained in the following terms: Large but typical values of $\sigma_{dB} \geq 8$dB result in a *ppr* = 0.5 contour "amoeba" that has "long legs." Provided the typical width ψ and average local node density are such that these pronounced "legs" are only just populated by neighboring nodes with a sufficiently high probability, the probabilistic localized routing algorithm will select such next hop neighbours consistently, resulting in significantly shorter end-to-end path hop counts. To support, clarify, and quantify this statement, further theoretical and simulation work is being pursued.

7.9 Conclusions

In this work we have shown that a meaningful assessment of a geographic, position-based routing protocol can only be achieved through careful incorporation of the physics of radio wave propagation in simulations. We demonstrate that UDG-based models, and models that fail to take into account the fully correlated spatial distribution of link reliability, significantly over-predict the end-to-end path hop count. The only geographic routing algorithm that has been found to be capable of exploiting the distant neighbor forwarding opportunities that occur in physically realistic radio wave propagation environments, and thus exploited end-to-end paths requiring fewer hops, is the localized probabilistic progress algorithm [15]. The corollary to this conclusion is that failing to adopt a physically accurate model for radio wave propagation can under certain circumstances ($\psi = 1$ and $\sigma_{dB} \geq 8$dB) produce pessimistic results on this protocol's performance in simulations.

We have provided a complete overview of the manner in which the physical radio environment needs to be modeled. Furthermore, we have identified three parameters that are required in order to meaningfully characterize position-based routing and provided a geometric interpretation for these:

- The angular spread of regions of higher than average range for a given probability of packet reception, ψ, which is determined by the shadow fading de-correlation length and the notional node transmission radius
- The extent to which such regions are pronounced, given by the shadowing standard deviation σ_{dB}
- The average local node density, κ, which ultimately determines the probability with which such regions are populated by neighboring nodes that are likely to receive packet transmissions successfully

References

1. Broch J, Maltz DA, Johnson DB, Hu YC, Jetcheva J (1998) A performance comparison of multi-hop wireless ad hoc network routing protocols. In: Proceedings of the 4th Annual ACM/IEEE International Conference on Mobile Computing and Networking, Dallas, TX, USA, pp 85–97
2. Takai M, Martin J, Bagrodia R (2001) Effects of wireless physical layer modeling in mobile ad hoc networks. In: Proceedings of the 2nd ACM International Symposium on Mobile Ad Hoc Networking & Computing, CA, pp 87–94
3. Qin L, Kunz T (2003) On-demand routing in MANETs: the impact of a realistic physical layer model. ADHOC-NOW, Montreal, Canada, pp 37–48
4. Konstantinou K, Kang S, Tzaras C (2007) A measurement-based model for mobile-to-mobile UMTS links. IEEE VTC Spring, April, pp 529–533
5. Gudmundson M (Nov 1991) Correlation model for shadow fading in mobile radio systems. Electron Lett 27(23): 2145–2146
6. Wang Z, Eustace K, Nix AR (January 2008) Joint shadowing process in urban peer-to-peer radio channels. IEEE Trans Vehicular Technol 57(1):52–64
7. Vaughan R, Bach Andersen J (2003) Channels, propagation and antennas for mobile communications. In: IEE electromagnetic waves series 50. IEE, London

8. Wang Z, EustaceK, Nix AR (2004) Statistical peer-to-peer channel models for outdoor urban environments at 2GHz and 5 GHz. IEEE VTC2004 Fall, Los Angeles, Oct 2004, pp 5101–5105
9. Press WH, Teukolsky SA, Vetterlink WT, Flannery BP (2007) Numerical recipes: the art of scientific computing, 3rd edn. Cambridge University Press, Cambridge, §2.9
10. Benedetto S, Biglieri E, Castellani V (1987) Digital transmission theory. Prentice-Hall, Englewood Cliffs, NJ, p 207
11. Zorzi M, Pupolin S (1994) Outage probability in multiple access packet radio networks in the presence of fading. IEEE Trans Vehicular Technol 43(3):604–610
12. Stüber GL (2000) Principles of mobile communication, 2nd edn. Springer, Berlin
13. Kuruvila J, Nayak A, Stojmenovic I (2005) Hop count optimal position based packet routing algorithms for ad hoc wireless networks with a realistic physical layer. IEEE J Selected Areas Commun 23(6):1267–1275
14. Finn GG (1987) Routing and addressing problems in large metropolitan-scale internetworks. ISI research report ISU/RR-87-180
15. Stojmenovic I, Nayak A, Kuruvila J (March 2005) Design guidelines for routing protocols in ad hoc and sensor networks with a realistic physical layer. IEEE Commun Mag 43(3):101–106
16. http://www.xbow.com/Products/Product_pdf_files/Wireless_pdf/IRIS_Datasheet.pdf
17. Li CP, Hsu WJ, Krishnamachari B, Helmy A (2005) A local metric for geographic routing with power control in wireless networks. Second Annual IEEE Communications Society Conference on Sensor and Ad Hoc Communications and Networks, IEEE SECON 2005, pp. 229–239, 26–29 September

Chapter 8
Medical Applications of Wireless Sensor Networks: Who-Did-What

Stanislava Stanković

Wireless Sensor Network (WSN) is a set of small, autonomous devices, working together to solve different problems. It is a relatively new technology, experiencing true expansion for the past decade.

Research in the field of nanostructures and sensors has brought real opportunities for development of WSNs. People have realized that integration of small and cheap microcontrollers with sensors can result in production of extremely useful devices, which can be used as an integral part of the sensor nets. These devices are called sensor nodes. Nodes are able to communicate with each other over different protocols. Studies in the field of communication protocols for wireless sensor networks are particularly interesting, and rely on various network topologies (Fig. 8.1). Issues, addressed by communication among nodes, include power management, data transfer, mobility patterns, etc.

Elements, capable of sensing different parameters, serve as data collectors in these networks. The gathered data can be wirelessly transfered to any external point, or simply interchanged among sensor nodes. The fact that data, sensed by sensors, can leave the network highlights the potential of WSNs, their usability in various fields.

8.1 WSN – Youth and History

As mentioned before, WSNs present new technology. History of this concept began at University of California (UC), Berkley, with Smart Dust project, which was funded by Defense Advanced Research Projects Agency (DARPA) [1]. The aim of this project was to develop self-organized, millimeter-scale hardware platform for distributed WSNs. Primarily, this was a military application which resulted in the development of relatively large sensor nodes. Later, miniaturization process brought much smaller devices, with solid sensing and communication capabilities.

S. Stanković (✉)
University of Belgrade
e-mail: saka58@gmail.com

L. Gavrilovska et al. (eds.), *Application and Multidisciplinary Aspects of Wireless Sensor Networks*, Computer Communications and Networks,
DOI 10.1007/978-1-84996-510-1_8,

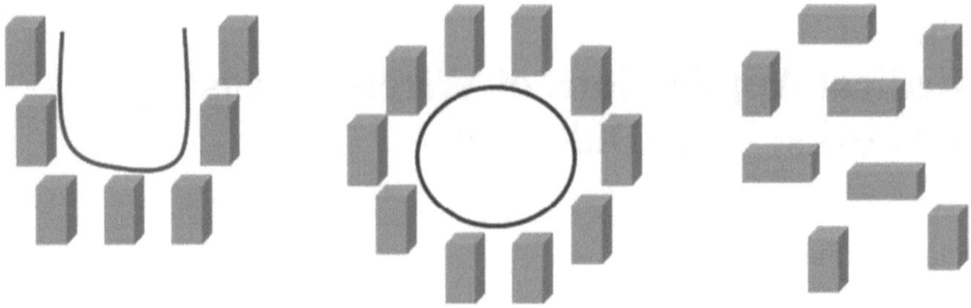

Fig. 8.1 Different topologies of WSNs

One of the key points in the history of wireless sensor networks was the implementation of energy-efficient software platform, Tiny OS, operating system, also developed at UC. Further development led to implementation of different software platforms for WSNs. This was well-explained in a previous chapter.

Soon, people understood that by putting sensor nodes to work together, infrastructure improvement and problem resolving can be achieved in different fields. The opinion, which was generally accepted, points on the low cost of this technology and its endless benefits. Today, sensor nets are used in agriculture, ecology, and tourism, but medicine is the area where they certainly meet the greatest potential.

8.2 WSN – Open Questions of Today's Healthcare System

Nowadays, healthcare system is highly complex. The list of elderly people and people in need of continuous care increases every day. Medical staff is faced with more and more challenges each year. This opens serious questions in the domain of interest, which must be answered in the best way possible. Problem solving must include detail analyses of the current state in order to form a good and functional system which resolves satisfactorily the number of issues. Wireless sensor networks can offer this kind of solution.

Patients in need of continuous monitoring or long-term care, meet a number of difficulties. Accessibility to medical experts, for these patients, is an issue as much as expenses. In case of accidents, the most important thing is to insure emergency help for persons with most urgent injuries.

Sensor nets can give a solution for the mentioned issues. This technology can improve both, the process of patient monitoring and the process of emergency response.

Healthcare system today meets accessibilty and mobility problems which imply a great deal of costs. The accessibilty problems refer to the absence of medical experts, in the sense of the right place and the right time. On the other hand, mobility problems address situations where persons have to stay in hospitals even when not neccessary, because of medical equipment dependencies. All these problems demand a completely new approach, a new concept of health management, which is offered by wireless sensor networks.

8.3 WSN – The Most Powerful "Weapon" of Modern Healthcare

A new concept of health management is the neccessary point which must be established in the near future. The good aspects of sensor networks will be discussed in the following paragraphs. Their potential of being the most powerful "weapon" of modern healthcare, will be explained.

Precise planning and adequate design of sensor net may lead to the development of a stable and functional system. Many advantages are discovered by putting sensor nodes to work together and continuously collect data. Nodes sense vital parameters (body temperature, pulse, pressure, oxygenation levels) and do some basic data processing. WSN also assumes that gathered information can be extracted from the network itself and sent to more powerful machines. This opens new views to the concept of healthcare.

Simple data sensing and processing is already one step forward, but data extraction enables us to track a patient's entire clinical history and to provide medical staff with information of any kind. We can use data mining in order to discover hidden facts about someone's condition. WSNs with attached data mining uncover a number of possibilities, from theoretical as well as practical aspects. Knowledge is power and in this case, power increases doctors' and nurses' efficiency and prevents mistakes.

It is extremely important to cover a great number of medical scenarios.

WSNs give us the opportunity to face challenges. This technology offers a solution for four important healthcare patterns:

- Pre-hospital care
- In-hospital care
- Post-hospital care
- Continuous care

Now, each one of the mentioned patterns will be further discussed.

Pre-hospital care refers to situations of mass accidents or any other kinds of emergency help situations, when physicians are faced with lack of time. For them, it would be very useful to have an insight into the real state of injured people, and sensor nets can provide this kind of control.

In-hospital care implies continuous monitoring of patients' vital parameters, and appropriate data transfer to persons in charge.

Post-hospital care includes post-surgery parameter tracking, or any other situations where people face illness recovery.

Continuous care involves all three previous patterns, but also refers to smart homes and elderly care.

Besides the mentioned facts, WSNs, as the "weapon" of healthcare, assume far more possibilities. For example, if we include implantable sensors, we get a completely new face of medicine. At this very point, the off-the-shelf market is not overfilled with these products, so this subject is only mentioned here, and may be considered in some future study.

We must be aware of the current state in the field of interest, in order to generate innovative ideas. The aim of this chapter is to uncover the state-of-the-art in medical applications based on WSNs, so in the following we will introduce a few large projects being developed in research sectors. Firstly, we will talk about important issues these applications are facing and the technologies in use. Projects include various scenarios which can be categorized in at least one of the four patterns emphasized earlier.

8.4 WSN – Existing Medical Applications (EXMAs)

The stress of this chapter is on an overview of medical applications based on WSNs, being developed in large research sectors. Projects we talk about, are simply unified as EXMAs, and further will be specified as such. EXMAs are recognized as a significant part of today's research activities in the area of medical applications of sensor networks. By highlighting the essence of these applications we understand the way WSNs can facilitate medical proficiency.

Firstly, we will indicate some important issues EXMAs are facing and which must be resolved during the implementation or deployment phase. Then, we will emphasize the technologies WSNs rely on and give a comparison of a few selected applications.

8.4.1 EXMAs – Important Issues

The main issues of medical applications based on wireless sensor networks can be divided into three categories (Fig. 8.2):

- **Technical issues**
- **Social issues**
- **Personal issues**

Three specified domains cover a remarkable number of problems. First one, the domain of technical issues, is common to all applications depending on WSNs. Social and personal issues are more related to health applications.

Now, we will explain all three types of EXMAs' issues. The interpretation itself aims to uncover things we need to have in mind, in case we want to design and implement a sensor network for some medical use.

Technical issues include power management, communication protocols, networks' reliability, and security.

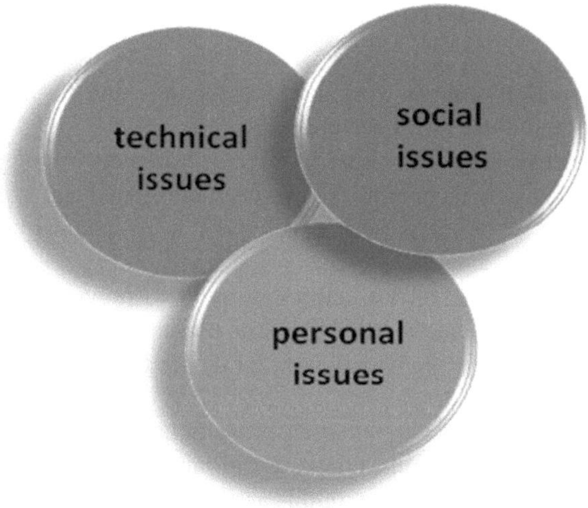

Fig. 8.2 Possible issues of EXMAs

Power management is the most important part of any WSNs' infrastructure, because of the fact that energy is limited. Data sampling and data processing are energy consuming, but the biggest energy part is consumed by data transfer. Designers of sensor nets must accurately define how the data will be transferred through the network, in order to conserve energy as much as possible.

Communication protocols are another type of technical issue. Well-structured protocols aim to resolve all the problems engineers are faced with. This includes power management, but also reliability and security, and other problems like congestion avoidance, load balancing, real-time services, etc. The stated concepts are real problems in the world of sensor nets, and improvement in this area is a crucial part of research.

Security could be referred as a technical, but also as a social issue. Data security and privacy must be provided in order to achieve acceptance of WSNs' technology. People are not willing to share their privacy, so different encryption techniques are used to secure data transfer through sensor networks. In case of medical applications, security of transferred data can be compared to doctors' ethic.

Besides security, **social issues** refer to general sensor nets' impact on social life. This technology must be well fitted into existing infrastructures. Earlier, we have mentioned that first designs of sensor nodes were quite large and that it was not satisfactory. New devices are small, but miniaturization process aims to produce even smaller and well-designed nodes. Nowadays, people carry a number of devices, mobile phones, i-pods, GPS devices, so the list expansion should not be a problem, as long as new devices fit the profile.

Personal issues have a physical and a psychic face. Sensor nets, in any way, may not harm patients' health. At this point, it is important to have in mind that WSNs aim to improve patients' lives and not to compromise them.

8.4.2 EXMAs – WSNs vs. Existing Infrastructures

Here, we will mention some networking infrastructures, and discuss about the hierarchy of WSNs and those infrastructures (Fig. 8.3).

The face of networking today is very different from networking just a of couple decades ago. Many new technologies have been developed and standardized. One of the advantages of wireless sensor networks is its compatibility with these technologies.

WSNs rely on existing infrastructures in order to provide real-time services. Radio frequency identification (RFID) is low-power technology with number of benefits. Today, RFID tags are widely used for device tracking, so they can be used to keep track of sensor nodes.

Locating is an essential part in some projects based on WSNs, for example, emergency response applications. Global Positioning System (GPS) can be identified as basic supporting technology in this area.

Data, collected by sensor nets, need to be transferred and presented in some way. Choosing the Internet as representation portal seems like the most appropriate solution. This can be done over General Packet Radio Service (GPRS). Internet connectivity can also be reached over wireless local area networks (WLANs/802.11). Representation of the same data on mobile phones can be achieved via Bluetooth technology.

We have introduced basic technologies that sensor networks rely on. Combining of different frameworks leads to deployment of different appliances of sensor nets, wireless personal area networks (WPANs), or wireless body area networks (WBANs).

8.4.3 EXMAs – State of the Art

Different projects, based on WSNs, are being developed, worldwide. These projects can mostly be referenced as academic researches, still finding their way into industry

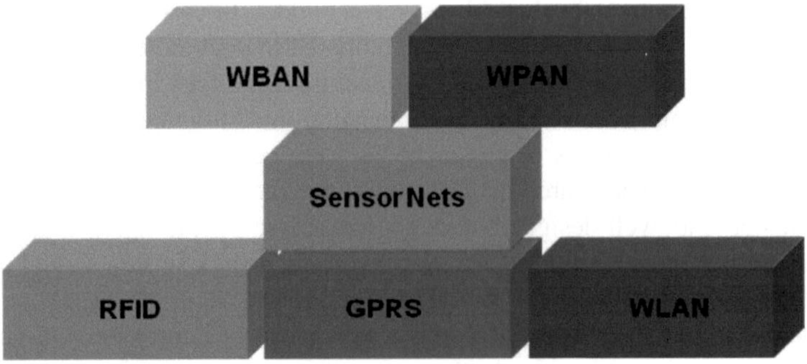

Fig. 8.3 Hierarchy of WSNs and existing infrastructures

sectors. Here, we will mention couple of projects with different interests, and give a corresponding comparison. The comparison will be done in terms of general description, hardware, software, and expected contribution. This overview aims to discover the progress which has been made in the area of EXMAs, for the past few years. The purpose of this preview is to provide persons willing to develop something similar, with sound fundamental knowledge.

8.4.3.1 UbiMon

UbiMon (ubiquitous monitoring environment for wearable and implantable sensors) is the architecture for distributed mobile monitoring, developed at Imperial College London. The aim of this system is to provide continuous management of patients under their natural physiological states so that transient but life-threatening abnormalities can be detected and predicted [2]. The use of wearable and implantable sensors is the area of research with a promising future, and the UbiMon system presents the framework for integration of these sensors into the body sensor network (BSN).

Hardware

The UbiMon hardware includes several components:

- BSN nodes
- Local processing unit (LPU)
- Central server (CS), and
- Workstation (WS)

BSN nodes provide patients' monitoring. In essence, these are miniaturized wireless intelligent modules based on MICA2 dot platform. Each node is integrated with a wearable or implantable sensor (ECG, SpO2, temperature) in order to track vital parameters of patients. Besides these, the BSN node also includes sensors for recognizing patients' activities and mobility patterns (accelerometer).

LPU, which can also be referenced as the base station, is a portable device (PDA or mobile phone). The device serves for sensor data gathering and processing. Data processing includes detection of abnormalities and providing the immediate warnings to patients. In addition, LPU also acts as a router between the BSN nodes and the central server via short-range wireless communication standards such as WIFI/Bluetooth or long-range mobile networks such as 3G/GPRS.

Central server receives real-time multisensory monitoring information from local processing unit, stores the data to the patient database, and performs long-term trend analysis. After the exhausting data analysis some physiological patterns can be extracted in order to enable the prediction of the patient's condition and prevention of potential life-threatening abnormalities.

Workstation is a portable device and serves as the monitoring terminal. It allows clinicians to analyze the patient data. Clinicians may see patients' entire clinical history over the terminal and perform real-time diagnoses.

Software

Software for the UbiMon architecture can be organized into five different modules. This categorization is based upon the existing hardware elements. Modules, which can be listed, are:

- BSN node software module
- LPU software module
- CS software module
- Patient database (PD)
- WS software module

The first module for the BSN node is TinyOS-based piece of software. It samples the sensor data and transfers them to local processing unit. High rate requirements of the sensor signals (ECG) resulted in TinyOS protocol and radio stack replacement with a lightweight protocol with special time division multiple access (TDMA).

LPU software communicates to BSN node via serial interface of the base station (MICA2 dot) to gather sensor data. The module is also responsible for displaying the gathered data and recognizing the patients' activities. Recognition is done based on the gathered context-aware sensor data. Another important software tool is WIFI/Bluetooth-based data transfer to central server.

CS module receives the data and stores them into patient database. This enables later purchasing of the patient clinical records, in any way.

WS software includes retrieval and proper representation of patients' data. Data retrieval is done over data extraction from database. Later, data are presented to clinicians through appropriate graphical user interface. This enables clinicians to perform successful patient management.

Contribution

The UbiMon system, besides continuous patients' monitoring in their natural environment, provides the architecture for collection, gathering, and analyzing data from a number of biosensors (wearable or implantable). Another significant aspect provided with this architecture is the context awareness which enables the capturing of any clinical relevant episode.

8.4.3.2 Satire

Satire (software architecture for smart attire) is a wearable personal monitoring service, developed through cooperation between University of Illinois and University of Virginia. It allows users to maintain a private searchable record of their daily activities as measured by motion and location sensors, which are two of the most popular sensing modalities in personal instrumentation [3]. The main idea is to deploy nonobtrusive wearable monitoring platform for data sensing, local storage, and occasional upload in order to satisfy medical, safety, personal, or entertainment reasons.

Hardware

Hardware components of the Satire architecture include

- MICAz sensor nodes
- Access motes
- PCs

MICAz motes were selected as suitable sensor nodes for the Satire prototype implementation. Other platforms, which also can be used, are MICA2dot, Tmote Sky, and Pluto nodes. MICAz nodes serve as **sensing**, local **storage**, reduced **processing**, and **communication** elements, with solid **energy supply**. Sensing implies usage of acceleration and GPS sensors, in order to capture user's mobility patterns. Data sampling requires 50 samples per second rate for 2-axis accelerometer, and 1 byte for sample representation. Node's flash memory can satisfy data recording requirement for about 6 h. Information processing in the nodes is minimal because of the unlimited data processing power, after the upload. Sensor nodes use wireless communication for data upload through the access motes. Power supply is in the form of two AA batteries per node which can be replaced with Lithium batteries to extend the node's lifetime.

The Satire system is not aiming to provide continuous care at all times. Continuity should be achieved by collecting information from different sources such as instrumentation in the person's home, office, and garments. The access motes are used to provide the necessary support for this feature. MICAz motes, with reduced capabilities, serve as base stations or access motes between the sensor nodes and PCs.

User's vicinity to access mote (connected to a PC) enables logged data upload to a private repository associated with a person. Further, the uploaded data is used for user's activities reconstruction. That also enables person's clinical history enhancement.

Software

The software architecture implements a heterogeneous system enabled for rapid changes' tracking due to new hardware facilities, new sensor modalities, and new appliances. In this case, the architecture of PC with attached access mote can be split in several layers:

- The application layer (4th)
- The parsing layer (3rd)
- The interpretation layer (2nd)
- The USB/Serial port layer (1st)

On the other hand, software on motes includes

- The application layer (4th)
- The sensor layer (3rd, 2nd, and 1st)

The application layer, for both cases, handles user interfaces for different applications. This module offers support for the extended functionalities over time which may arise, and also supports different storage modalities: databases and web-servers.

Depending on the expansion of sensor modalities, software prerequisites must be met and it is done in form of the parsing layer. The parsing layer enables additional flexibility of including new sensor into the system by processing the raw data generated by sensors.

The interpretation layer intends to handle different algorithms for processed data interpretation.

For motes, the sensor layer includes three sub-layers. Firstly, there are sensor-specific protocols which serve as filters. The filters implement data processing algorithms, for example, in terms of energy saving, stillness detection. Under this, the TinyOS and fundamental protocols for system's working are set. Protocols refer to data synchronization, data logging, and upload.

At the end, on the motes' side, layer 1 is UART/RF-Radio-based. It communicates to the USB/Serial port layer on the other side and enables data transfer.

Contribution

Major contribution of the Satire system is user transparency. Project's goal includes the implementation of the service which provides information collection, storage, and uploads without explicit input or maintenance required from the user. Satire presents new wearable platform limited by the currently available off-the-shelf components.

8.4.3.3 Alarm-Net

Alarm-Net is a wireless sensor network for assisted living and residential monitoring being developed on University of Virginia. It integrates environmental and physiological sensors in a scalable, heterogeneous architecture [4]. The focus of this network is to provide smart environment and to secure patients' monitoring on daily basis. This project aims to facilitate the concept of health care by including informal caregivers, such as family, friends, and patients' themselves in the process. Early detection and prevention of diseases insures cost-reducing with simultaneous care quality maintenance.

Hardware

The Alarm-Net architecture includes heterogeneous devices:

- MICAz sensor nodes
- Stargate gateways
- iPAQ PDAs
- PCs

MICAz nodes are integrated with infrared motion and dust sensors, and also temperature, light, pulse, and blood oxygenation sensors are present. Sensors for temperature, dust, motion, and light are referenced as **emplaced sensors**. These devices are deployed in the living space to sense environmental quality or conditions. Pulse and blood oxygenation sensors form **body networks** which provide physiological sensing and activity classification.

Stargate gateways are embedded platforms serving as gateways between the wireless sensor and IP networks. These nodes allow user interfaces to connect, authenticate, and interact with the system.

PDAs and PCs form **back-end** systems which provide online analysis of sensor data and long-term storage of system configuration, user information, privacy policies, and audit records. These components also serve as **user interfaces** which allow any legitimate recipient of the system to query sensor data, depending on enforced privacy policies.

Software

Software components of the Alarm-Net system are

- TinyOS query processor and motes' security modules (SecureComm)
- AlarmGate
- Java resident monitoring and sensor data querying applications
- Circadian activity rhythm (CAR) analysis program

TinyOS query processor serves as simple data sampler and limited data processor. Security modules enable communication between motes and stargates. Basic component is the implementation of SecureComm protocol, a hardware-accelerated secure messaging protocol.

AlarmGate is an embedded Java application which runs on stargate gateway. It can be referenced as application level gateway between the wireless and IP networks and serves for managing power, privacy, security, queries, and client connections.

Java resident monitoring and sensor data querying applications run on PDAs and PCs. These applications present and store collected sensor data.

A CAR analysis program processes stored data, learns individual behavior patterns, and provides support for context-aware power management and privacy by putting some feedback into the network.

Contribution

Contributions of the Alarm-Net system include development of scalable, heterogeneous network, based on custom hardware components; implementation of context-aware protocols, informed by CAR analysis, which enable smart power management and dynamic alert-driven privacy depending on individual's activity patterns;

design of SecureComm messaging protocol which supports user-based decisions for selection of security policies.

8.4.3.4 BikeNet

BikeNet is an extensible mobile sensing system, being developed through collaboration of Dartmouth College and Columbia University, for cyclist experience mapping leveraging opportunistic sensor networking principles and techniques. It represents a multifaceted sensing system and explores personal, bicycle, and environmental sensing using dynamically role-assigned bike area networking [5].

Hardware

The BikeNet architecture includes

- Moteiv Tmote Invent sensor nodes
- Nokia N80 mobile phones
- PCs and servers

These devices are organized in three tiers: the **mobile sensor tier**, the **sensor access point (SAP) tier**, and the **back end server tier**.

The mobile sensor tier incorporates a number of bicycle-mounted and human-mounted Tmote Invent platforms. Tmote nodes constitute a BAN. Platform native sensors in use are: a two axis accelerometer, a temperature sensor, a light sensor, and a microphone. Other sensors additionally attached to the motes include: a magnet-triggered reed relay for angular velocity of the wheel and pedal and distance measurements, a dual axis magneto-inductive sensor (Honeywell HMC1052L) for measurement of direction and deviation with respect to Earth's magnetic field, a Garmin Etrex 12 channel GPS unit for absolute time and location measurements, a Telaire 7001 CO_2/Temperature Monitor for CO_2 measurements, an ArcherKit Biofeedback Monitor for measurement of the galvanic skin response of the cyclist.

The SAP tier serves as gateway between the mobile sensor tier and the back end server tier. SAPs can be static and mobile. The static ones are wired directly to the Internet. In essence, static SAP is a Tmote Invent plugged into USB port of an Aruba AP-70 IEEE 802.11a/b/g access point. The mobile SAP is implemented using a Bluetooth enabled Nokia N80. Between its back end interface and the back end servers SAP uses GPRS/GSM communication, in case of mobile SAP, and TCP/IP communication, in case of static SAP.

The back end server tier includes power, high performance servers with large storage capabilities.

Software

Software components for the BikeNet network are organized versus isolated hardware tiers. Here, we will mention software layers of the appropriate hardware components.

Motes' applications insure intra- and inter-BAN management, data exchange services, and support for BikeNet roles' completion.

SAPs' applications enable two-side communication, node-based through data exchange services and server-based through query management. These applications also include ground truth sensing module, which acts like sensor data filter. This assumes that even SAPs have some sensors attached, in order to provide high data fidelity.

Applications, running on servers, gather a couple of modules which enable server-SAP communication, sensor data visualization, sensor data storing (over repositories and databases), and sensor data mining. Module which is referenced as query submission portal enables graphical presentation of cyclists' data on the web.

Contribution

Contributions of the BikeNet system include cyclist performance and fitness management by collecting data about cyclist's current and average speed, distance travelled, calories burned, along with time and location metadata; environment and experience mapping by emphasizing the healthiness levels in terms of pollution, allergens, noise, and terrain's roughness; long-term performance trend analysis which enables the upload of data traces into personal repository and selective data sharing; data collection and local presentation customized by cyclists; data query and remote presentation on the web-based portal; disconnected operations which refer to opportunistic networking paradigm, or the fact that data uploading happens according to the opportunities that arise as a result of cyclists' uncontrolled mobility.

8.5 WSN EXMAs – A Conclusion

To conclude, we can emphasize the expansion of WSNs for the past few years. This certainly is a rising technology. As stated before, a great advantage of sensor nets is their compatibility to existing infrastructures. Another advantage lies in the fact that these networks meet a number of completely different applications, agricultural, ecologic, and especially medical. Miniaturization of sensor nodes and overcoming of demonstrated issues will bring more sophisticated solutions and applications.

8.5.1 WSN EXMAs – Failure vs. Success

With all new technologies we face with problem of failure or success. Despite the promising features, risks of failure are always present. In order to overcome the problem we have to pass through some necessary points. First of all, we have to start with theoretical basses' learning, and then to do some practical work in order to gain knowledge and to improve the state of the art.

8.5.2 WSN EXMAs – The Ultimate Aim

Existing medical applications based on wireless sensor networks are still research projects with good potential for utilization. A great number of medical scenarios are being covered with these applications and that have opened a wide spectrum of benefits for caregivers. They should be able to provide the necessary care whether in emergency situations, or in some hospital environments, or in the form of smart homes. A lot of effort needs to be done before the EXMAs' ultimate aim is reached. In this case, the ultimate goal implies the improvement of human lives. For that end, from this very point we can say that the future of WSNs and their medical appliances looks highly promising.

References

1. Yacoub M (foreword), Yang G-Z (ed) (2006) Body sensor networks. Springer-Verlag, London, pp 24–26
2. Yang GZ et al. (2004) Ubiquitous monitoring environment for wearable and implantable sensors. In: International conference on ubiquitous computing (UbiComp), Tokyo, Japan, 11–14 September 2004
3. Ganti RK, Jayachandran P, Abdelzaher TF, Stankovic JA (2006) SATIRE: a software architecture for smart AtTIRE. In: International conference on mobile systems, applications, and services (MobiSys), Uppsala, Sweden, 19–22 June 2006
4. Wood A, Virone G, Doan T, Cao Q, Selavo L, Wu Y, Fang L, He Z, Lin S, Stankovic J (2006) ALARM-NET: wireless sensor networks for assisted-living and residential monitoring. Technical Report CS-2006–11, Department of Computer Science, University of Virginia, 2006
5. Eisenman SB, Lane ND, Miluzzo E, Peterson RA, Ahn G-S, Campbell AT (2007) BikeNet: a mobile sensing system for cyclist experience mapping. ACM SenSys, Newyork, 2007

Chapter 9
Optimized Positioning of ECG Electrodes for WSN Applications

Ivan Tomašić and Roman Trobec

9.1 Introduction

Electrocardiography has its foundations in the recording of the electrical activity of the heart over time, using electrodes placed on the skin. Two electrodes measure the electrical potential differences on the body surface induced by the electrical currents through the heart muscle and surrounding conductive tissues. The method is simple and noninvasive and, therefore, widely used for diagnostic purposes in cardiology. The initial breakthrough in recording the electrical activity of the heart came from Willem Einthoven at the beginning of the twentieth century. He was the first to identify the various deflections, assigning to them the letters P, Q, R, S, and T, and described the electrocardiographic features of a number of cardiovascular disorders. In 1924, he was awarded the Nobel Prize in Medicine for his discovery. Since Einthoven's time there have been many advances in electrocardiography. Over the years, 12-lead ECG (electrocardiogram) became the gold standard, with its diagnostic foundation recognized by most cardiologists.

The conventional 12-lead ECG is obtained from ten electrodes placed strategically on the patient's body. A combination of signals measured between certain electrodes, with the introduction of a reference potential point, enables the heart to be "observed" in horizontal and vertical planes from different "view" angles. A systematic and intuitive interpretation of the measured results is possible, and is often supported by automatic rule-based diagnostics.

The limited communication bandwidth in body wireless sensor networks (BWSNs) motivates the optimization of electrocardiographic recording in terms of the number of body electrodes. Minimization of the required distance between

I. Tomašić (✉)
Department of Communications Systems, Jozef Stefan institute,
Jamavo 39, 1000 Ljubljana, Slovenia
e-mail: ivan.tomasic@ijs.si

R. Trobec
Department of Communications Systems, Jozef Stefan institute,
Jamavo 39, 1000 Ljubljana, Slovenia

L. Gavrilovska et al. (eds.), *Application and Multidisciplinary Aspects of Wireless Sensor Networks*, Computer Communications and Networks,
DOI 10.1007/978-1-84996-510-1_9,

electrodes, and improvement of wearing comfort of the investigated person are also important issues. They are, however, in partial contradiction with the standard 12-lead ECG, which exploits distant electrodes positioned on arms and legs. The number of electrodes can be reduced and their optimal positioning found that differs from the standard 12-lead placing. The measurements from reduced electrode sets can serve for the reconstruction of the standard 12-lead ECG, which is directly applicable to the current medical knowledge. Because the 12-lead ECG contains redundant information, a technically more legitimate approach would be to develop adapted diagnostic rules based on the measurements obtained by the reduced number of ECG electrodes that still carry the same information about the heart activities, but such an approach would require too much time for new experiments, validations, and education.

Wireless technology can be introduced into ECG measurements on different levels and with different approaches. For example, a standard wired ECG device can have wireless access to the medical center, which enables remote monitoring. Measurements from the wired electrodes can be assembled in an on-body subsystem equipped with a wireless transmitter that enables wireless data transfer from the monitored subject to the diagnostic ECG device, enabling much more freedom of movement for the monitored subject. Finally, a wireless electrode (WE) can be implemented that enables the minimal use of wires on the body and, consequently, the maximal wearing comfort. Such a solution could also minimize the disturbing signals generated in the musculature that are considered as noise in the case of ECG measurements.

All these options for introducing wireless technology for ECG measurements can be combined to provide a fully wireless system. However, for the purpose of this chapter, we will focus just on wireless electrodes, emphasizing two important issues. First, the wireless electrode enables the measurement and transmission of only local potential differences; therefore, we are limited to differential (bipolar) measurements. Second, limitations of transmission bandwidth and power supply require a significant reduction of the number of wireless electrodes, raising the question about their minimal number required for an accurate reconstruction of the standard 12-lead ECG with the adequate diagnostic information.

We have investigated in more detail the approach based on a reduced number of electrodes and reconstruction of the standard 12-lead ECG. The minimal number of electrodes has been determined, and their positions and distances optimized for each investigated person. Personalized transformation matrices have been obtained using multichannel ECG (MECG) measurements performed initially on each user of the wireless ECG. Multivariate statistical methods, such as principal component analysis (PCA) and regression analysis, have been applied on 31-channel measurements for selecting the optimal positions of ECG wireless electrodes. Personalized transformation matrices have been calculated for the reconstruction of 12-lead ECG with minimum loss of diagnostic information. It has been shown that a linear combination of only three independent potentials from wireless electrodes suffices for an accurate reconstruction – significantly less than the ten independent potentials needed by standard 12-lead ECG.

A wireless electrode is composed of two skin electrodes connected by a conducting wire, with appropriate electronics for digitalization and transmission of the measurements. A small set of electric potentials from wireless electrodes can be used in the reconstruction of the standard 12-lead ECG. Wireless electrodes can constitute a BWSN that eliminates the need for additional wires and therefore improves the applicability of ECG device technology. Wireless technology is applicable in clinical ECG devices but is maybe even more promising for personal and mobile ECG devices, because of increased mobility and comfort.

Several research and experimental projects have shown the feasibility of such an approach [1]. We have shown that the accuracy of the reconstructed ECG can even be improved with a personalized approach. Our results show that the number of wireless electrodes can be reduced to three, if positioned optimally for each investigated subject. Such a positioning contributes to the diagnostic information and therefore increases the reliability of the reconstructed standard 12-lead ECG.

Advances in the area of computerized electrocardiology should lead to considerably improved ECG systems in the near future [1]. A particular area of recent interest is the use of limited lead systems that obtain ECG information from just a few recording sites. Such approaches involve the application of a transformation to yield the more familiar 12-lead ECG. Several approaches have been proposed for reducing the number of electrodes. The most popular is the EASI lead system [2], which is commercially available from Philips Medical Systems. This system consists of three bipolar leads while others, like Wei's [3], Drew's [4], and Nelwan's [5], use a subset of the conventional 12-lead ECG set. Some of the leads used in these systems are bipolar and some unipolar. Detailed descriptions of the leads obtained from wireless electrodes, their positioning, and the principles for reconstructing 12-lead ECG, are given in the following sections.

The methodology for the reconstruction of standard 12-lead ECG, which is proposed in this chapter, was supported by multichannel ECG (MECG) measurements that were obtained by 31 or 35 ECG electrodes placed on the front and back part of the human torso in positions that were determined by an orthogonal mesh [6]. The distances between the neighboring electrodes in the mesh were 5–7 cm. The MECG provides 31 or 35 independent ECG potentials referenced to a common potential, while potential differences between any two electrodes can be regarded as bipolar leads. There are a lot of possible bipolar leads, but we addressed only the pairs that are composed of neighboring electrodes that are close enough. Their potential is electrically equivalent to the potential obtained from a wireless electrode of the same dimension as the distance between neighboring MECG electrodes. In this way we can emulate different numbers and positions of the wireless electrodes in the research and development of the reconstructed 12-lead ECG. Such a selection enables a straightforward design of strip-like wireless electrodes connected in a BWSN, which is our final goal [7].

The rest of this chapter is organized as follows. The 12-lead ECG is introduced in the next section with some basic terminology used in electrocardiography. An overview of different lead sets used for ECG recording is presented in Section 9.3.

We put more emphasis on the various existing approaches to the reduction of lead set. In Section 9.4, further details are presented on determining the minimal number of electrodes and their optimization regarding positions and distances. Multivariate statistical methods, i.e., PCA and regression analysis, are applied to the 31-channel MECG measurement for selecting optimal electrodes and calculating a personalized transformation matrix for the reconstruction of 12-lead ECG with minimum loss of diagnostic information. In Section 9.5, the proposed methodology is validated for a real test case. The chapter concludes with a critical overview of the presented results and required work for future wireless ECG systems.

9.2 Electrodes and Lead Types

The surface electrodes are electrical contacts positioned selectively on predefined places on the body surface. The electrical potential difference between two body surface electrodes is a consequence of the electrical currents through the heart muscle and surrounding conductive tissues. The measured electric potentials from different positions are strongly related to the diagnostic state of the heart, to be evaluated in different parts of the heart muscle. The meaning of "lead" is historically a combination of electrode potentials. Each lead represents an electrical axis onto which the electrical activity of the heart is projected. One may consider each lead to represent a different spatial perspective of the heart's electrical activity. There are two types of leads: unipolar and bipolar. A unipolar lead is, for example, the potential difference between an electrode usually placed on the chest and a reference potential, while a bipolar lead is a potential difference between two electrodes. We will provide a concrete example from the standard 12-lead ECG system.

9.2.1 Standard 12-Lead ECG

The standard 12-lead ECG uses ten electrodes, six of them, V1–V6, are precordial electrodes placed on the chest, the next three are limb electrodes placed on the right (RA) and left arms (LA) and left leg (LL), and the last is ground electrode, placed on the right leg (RL) [8]. Precordial chest electrodes are used to form six unipolar leads while the electrodes at the extremities form three bipolar limb leads, I, II, and III. Augmented limb leads aVR, aVL, aVF, derived from I, II, and III, are considered to be unipolar. V1–V6, I, II, III, and aVR, aVL, aVF, constitute the 12 leads of the standard ECG. Figures 9.1 and 9.2 indicate the standard electrode positioning in the 12-lead ECG.

Precordial leads are generated as electrical potential differences between precordial electrodes and Wilson's central terminal potential, which is the negative "electrode" of all precordial leads, and these leads are therefore considered to be

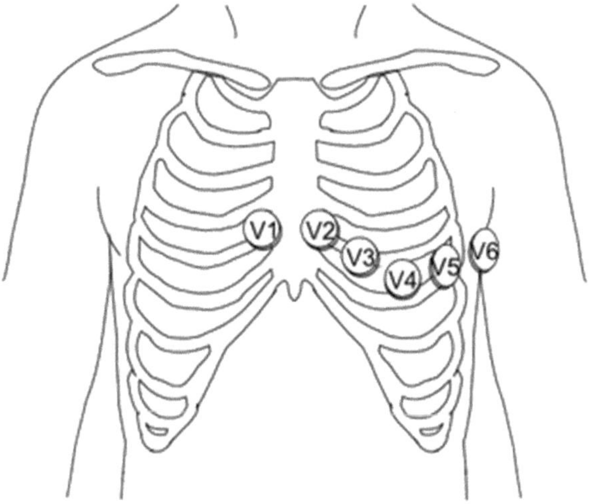

Fig. 9.1 Positioning of precordial electrodes of 12-lead electrocardiogram (ECG)

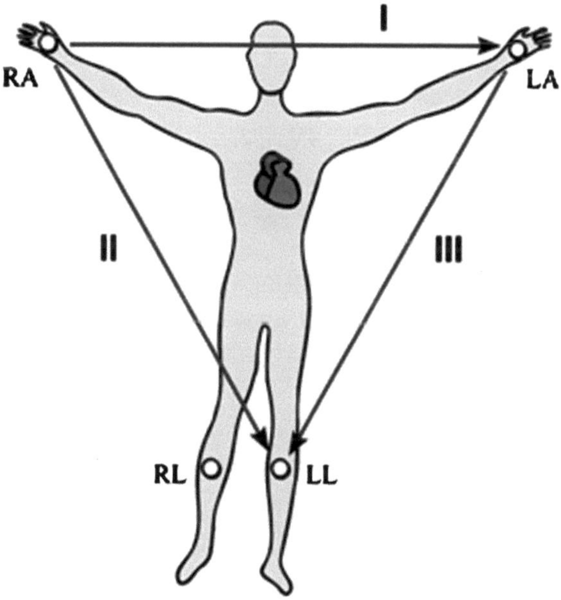

Fig. 9.2 Einthoven triangle

unipolar. The precordial leads enable the observation of the heart in the horizontal plane, across the heart center, from angles covered by V6–V1 in steps of approximately 15°.

Limb leads are traditionally denoted by I, II, and III, and obtained as electrical potential differences from the pairs of limb electrodes,

$$\mathrm{I} = \mathrm{V_{LA}} - \mathrm{V_{RA}},\ \mathrm{II} = \mathrm{V_{LL}} - \mathrm{V_{RA}},\ \mathrm{III} = \mathrm{V_{LL}} - \mathrm{V_{LA}} \tag{9.1}$$

where V_{LA}, V_{RA}, and V_{LL} denote the potentials on the limb electrodes. The Wilson's central terminal [8] is defined by the average potential from the limb electrodes: $(\mathrm{V_{LA}} + \mathrm{V_{RA}} + \mathrm{V_{LL}})/3$. The limb leads can be visualized by a well-known Einthoven's triangle (Fig. 9.2) with a Wilson's central terminal "electrode" at its center. In this way, the limb leads enable the heart to be observed in the vertical plane from angles 0°, 60°, and 120°.

The observation angle may be refined with −30°, 90°, and −150° angles (Fig. 9.3) obtained from augmented limb leads, aVL, aVF, and aVR derived from the combination of limb leads as:

$$\mathrm{aVR} = -\frac{\mathrm{I} + \mathrm{II}}{2},\ \mathrm{aVL} = \mathrm{I} - \frac{\mathrm{II}}{2}, \mathrm{aVF} = \mathrm{II} - \mathrm{I}/2 \tag{9.2}$$

Alternatively, augmented limb leads can be derived directly from augmented differences between limb electrodes and Wilson's central terminal potentials. For example, for aVR:

$$\begin{aligned}\mathrm{aVR} &= \frac{3}{2}\left(\mathrm{V_{RA}} - \frac{(\mathrm{V_{LA}} + \mathrm{V_{RA}} + \mathrm{V_{LL}})}{3}\right) \\ &= \mathrm{V_{RA}} - (\mathrm{V_{LA}} + \mathrm{V_{LL}})/2.\end{aligned} \tag{9.3}$$

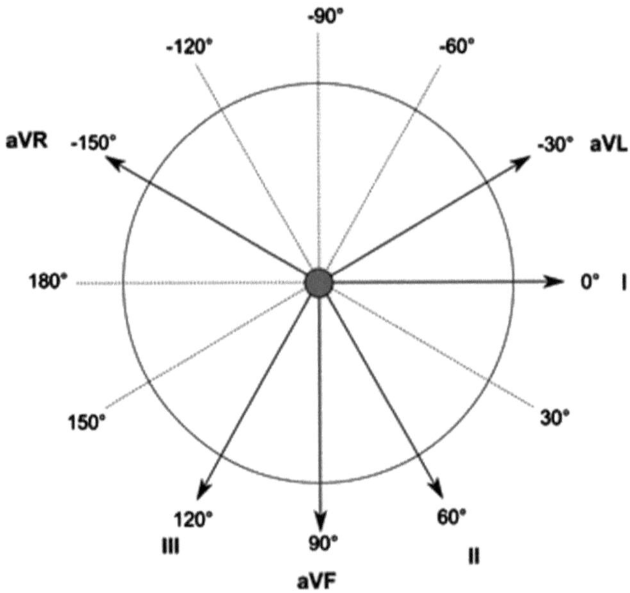

Fig. 9.3 Limb leads and their observation angles of the heart in the vertical plane

Rearranging in the expected form and using the definitions of limb leads we obtain:

$$\begin{aligned} aVR &= -\frac{1}{2}\left(-2V_{RA} + V_{LA} + V_{LL}\right) \\ &= -\frac{1}{2}\left(V_{LA} - V_{RA} + V_{LL} - V_{RA}\right) \\ &= -(I + II)/2, \end{aligned} \tag{9.4}$$

as expected.

9.2.2 *Differential Lead*

In the rest of this chapter we will use mainly measurements from the multichannel ECG, where each channel represents a unipolar lead, because its electrode potential is referenced to the Wilson's central terminal potential. The differential lead (DL) is defined as a difference between two MECG leads, which is equivalent to the potential difference between the two electrodes.

The wireless electrode (WE) is a device composed of two close body surface electrodes connected by a conducting wire, with appropriate electronics for digitalization and transmission of the measured data. Its measurements are bipolar and can be emulated by differential leads. The WE can be implemented in an adhesive strip-like form including electrodes, signal conditioning electronics, radio, and power supply. In some cases two WEs can share a common electrode, resulting in reduced number of required electrodes. The WE is constrained significantly by the transmission range and power consumption [9] on the one hand, and by the signal-to-noise ratio, because of the proximity of electrodes [10], on the other. Several WEs can be connected in a BWSN which eliminates the need for additional wires and therefore improves the applicability of the proposed technology [11]. Besides WSN use in clinical ECG devices, the proposed approach is also useful for personal and mobile ECG devices because of increased mobility and comfort due to the absence of wires [12].

The number of used WEs should be as small as possible because of limited data bandwidth and in order to limit power consumption. Most of the eventually redundant information needs to be extracted. We will focus in the following on methods for reducing the number of WEs and attempt to optimize their positions on the monitored person.

9.3 Alternative Methods for Positioning ECG Electrodes

Before explaining the different approaches to reducing the number of leads, we will illustrate some alternative positioning of electrodes currently used in electrocardiography.

9.3.1 Holters

Digital Holter monitors are ECG devices with reduced numbers of electrodes. They are becoming more and more popular for long-term personal ECG diagnostics. The number of channels, and therefore the number of electrodes and their location, differs between the numerous Holter producers since there is still no standardized solution for the best positioning (Table 9.1). The electrode location of the Breamar DL800 Holter monitor is shown in Fig. 9.4 [13].

9.3.2 Multichannel ECG

Systems with a lot of electrodes, at the other extreme, are multichannel ECGs exploiting many more electrodes than are needed for the standard 12-lead ECG.

Table 9.1 Electrode positioning of the Breamar DL800 Holter monitor

Channel	Placement
1−	Right manubrial border of the Sternum
1+	Left anterior axillary line sixth rib
2−	Left manubrial border of the Sternum
2+	Approximately 1 in right of Xiphoid
3−	Center of the manubrium
3+	Mid-clavicular line sixth rib.
Ground	Lower right rib margin over bone

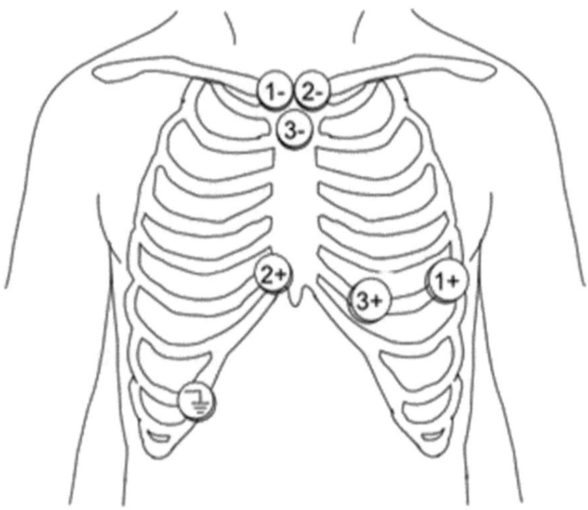

Fig. 9.4 An example of positioning for Holter's seven electrodes

Currently, MECGs are mostly experimental devices with no accepted standard as to the number of electrodes and their location. The number of MECG electrodes ranges from 10 to 300. Their placing is mostly based on the equidistant four neighbors mesh. We have developed a custom MECG [6] with either 31 or 35 electrodes, placed as shown in Figs. 9.5 and 9.6 and all referenced to the Wilson's central terminal potential. The two sets of placements are similar; however, that in Fig. 9.6 was adopted for patients with limited mobility. High-precision input amplifiers and filters are used for conditioning the input signals, and a 16-bit A/D convertor was applied for the conversion of input analogue signals to digital data. The sampling rate of each channel is 1,000 Hz. The battery-powered measuring device is connected optically to the computer system, which is responsible for data acquisition, visualization, and analysis.

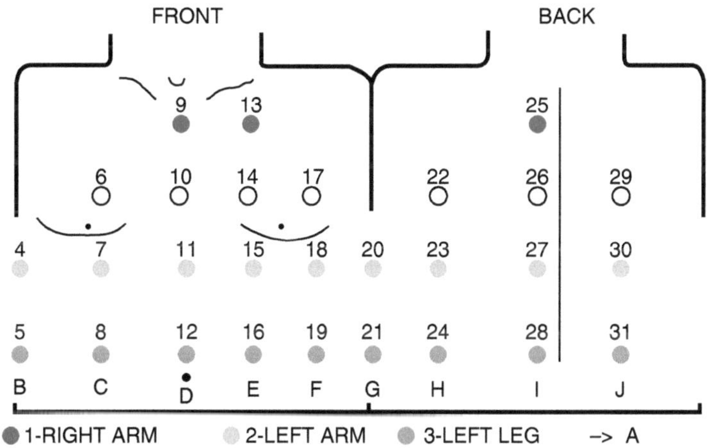

Fig. 9.5 Multichannel ECG (MECG) positioning of 31 electrodes

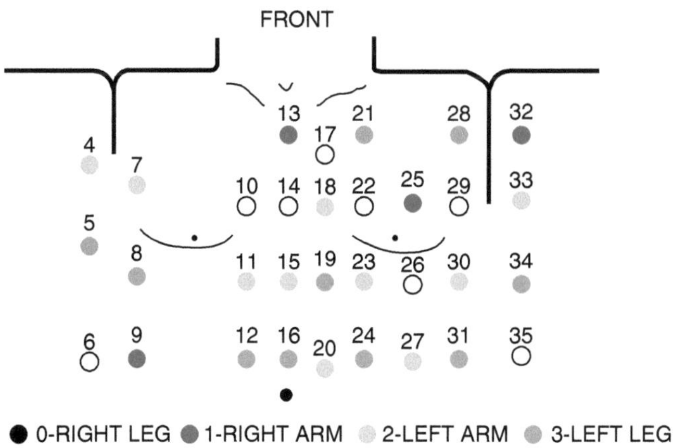

Fig. 9.6 Multichannel ECG (MECG) positioning of 35 electrodes

9.3.3 Lead Systems for Reconstructing 12-Lead ECG

Different dispositions of the electrodes from the reduced electrode sets have been adopted [14]. Most of them are based on the assumption that a single, time-dependent dipole can be used to represent the heart activity [15]. However, heart activity can only be modeled accurately by several dipoles that change their spatial and time states. Therefore, exact reconstruction of a measurement is, theoretically, not possible with a smaller number of electrodes.

The EASI lead system [2] consists of four signal electrodes plus ground electrode (Fig. 9.7). The four recording sites result in three independent bipolar measurements, E-S, A-S, and A-I. The 12-lead ECG is derived from a linear combination of these three measurements, using optimized, fixed coefficients. The locations of leads were chosen based on good anatomical landmarks allowing accurate electrode location, good signal-to-noise ratio, and convenience for patients and technicians.

As for most limited lead systems, the main advantage of the EASI system is that the 12-lead ECG can be reconstructed from less than the ten recording electrodes previously required. A further advantage is its reduced susceptibility to motion artifacts, since the recording sites are located exclusively on the torso, away from the extremities. Accurate positioning of the electrodes is also made easier, since recording locations are at prominent anatomical landmarks. This is in contrast to the precordial leads of the standard 12-lead ECG, which are often difficult to locate accurately, particularly in women and children.

Nelwan et al. [5] investigated how well some absent precordial leads could be reconstructed from the remaining leads of the standard 12-lead ECG. Sixty-three different subsets of precordial leads were investigated with one or more leads removed. For each subset, transformation coefficients were developed from which

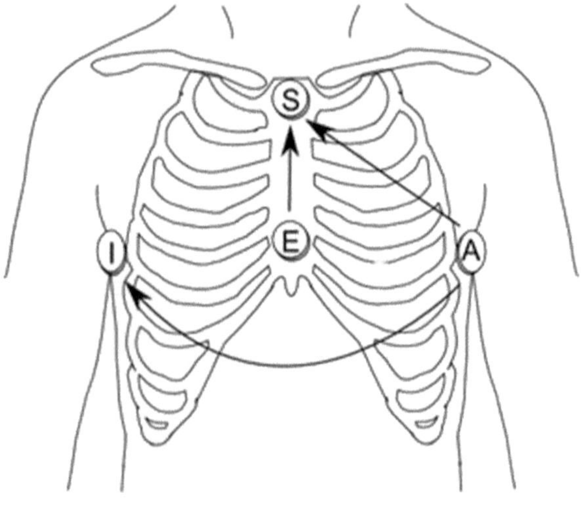

Fig. 9.7 EASI lead positioning

Table 9.2 Main characteristics of limited lead systems

Lead System Name	Year	Leads	Evaluation
EASI, Dower at al.	1988	Three bipolar	Numerous studies
Wei	2002	I, II, V1, V6	113 Subjects
Drew et al.	2002	I, II, V1, V5	649 Subjects
Robertson et al.	2002	RV2, RV4, RV6, and RV1, RV2, RV4	64 Subjects
Nelwan et al.	2004	I, II, V2, V5	38 Subjects

the 12-lead ECG could be accurately reconstructed when as many as four precordial leads were removed. In another study [16], 24-h ECG recordings were analyzed. Reconstruction of up to four leads was shown to be possible. In a further study [17], 12-lead ECG reconstructed from a system consisting of leads I, II, V2, and V5 was compared with the standard 12-lead ECG. This new lead system reconstructed the traditional ECG even more accurately, in terms of root means square error, than the EASI lead system.

Wei [3] also used a subset of a standard set of leads in conventional 12-lead ECG for reconstructing 12-lead ECG. Leads I, II, V1, and V6 were investigated for their reconstruction ability. Both waveform shape and diagnostic information were compared between the reconstructed and original signals. Wei concluded that this system had a significant advantage over the EASI lead system. One of the advantages was that several of the 12-lead signals were measured directly and therefore needed no reconstruction.

Drew et al. [4] used a lead set that differed from Wei's lead set in only one lead, using V5 instead of V6, making it a lead set of: I, II, V1, and V5. This system was evaluated on 649 subjects. The authors concluded that 12-lead ECG derived from this set of leads was valuable for diagnosing multiple abnormalities, including wide-QRS-complexes, tachycardia, and acute myocardial ischemia/infarction.

Robertson et al. [18] introduced two limited lead sets consisting of bipolar leads. The first used V2, V4, and V6 precordial leads, which were measured against right-arm potential. The other bipolar lead system also used right arm as one side of the bipolar lead, but with V1, V2, and V4 precordial leads.

Most of the listed lead systems have been evaluated in terms of their diagnostic value. However, because of huge number of possible abnormalities it was not proven that all abnormalities can be detected with the same accuracy as with the standard 12-lead ECG. The main characteristics of the limited lead systems are summarized in Table 9.2.

The underlying principle of the reconstruction method for all limited lead systems is the transformation of information recorded from a limited lead system to the target lead system [1]. A set of transformation coefficients is defined on the basis of optimization and statistical approaches. The transformation that is most often used and proven to be effective and simple is linear transformation:

$$S = \alpha_1 L_1 + \alpha_2 L_2 + \ldots + \alpha_i L_i + \ldots + \alpha_p L_p \tag{9.5}$$

where S is the reconstructed lead, α_i are constant coefficients, and L_i are measured leads from a limited lead system composed of p leads. The above equation can be written in a matrix form for r reconstructed leads:

$$S = A \cdot L,$$

where A is an $[r \times p]$ matrix, L and S are vectors of p input leads from limited system, and r output reconstructed leads, respectively. The transformation coefficients are time-independent and the same for all investigated subjects.

9.4 New Methodology for Wireless Electrode Positioning and Reconstruction of 12-Lead ECG

9.4.1 Input Data

The input data set should be as large as possible, i.e., as many test electrodes as possible should be used. Larger numbers of test electrodes increase the chances of better selection of an optimal subset of electrodes. Our input data sets were measurements from 31-channel MECG, 10 s long (10,000 samples per channel), either consecutive from the same subject or single measurements from a number of subjects. Note that an MECG measurement has enough leads, placed at appropriate positions, to generate the target 12-lead ECG, which is the same as the one obtained by a standard measurement.

For detailed verification of the selection algorithm it is useful to have several test MECG measurements from the same patient. The first MECG measurement is used to calculate the transformation coefficients required by the reconstruction algorithm, and to select a set of the best electrodes for the reconstruction of the standard 12-lead ECG. Note that the remaining MECGs are, in fact, not known in real situations, where only measurements from a limited number of leads will be available. But if available, they can be used for the evaluation of the obtained reconstruction parameters – differential lead positions and transformation coefficients.

In this way, validation of the reconstruction algorithm is possible using different measurements from the same subject, taken on different occasions, for example, before and after heart surgery. If only a single test MECG measurement is available, we can still evaluate the derived reconstruction coefficients, with some confidence. In this work we evaluated the algorithms for selecting optimal differential leads with 31-channel test MECG measurement that contains, beside the normal sinus rhythm, a single supraventricular extrasystole. Such a reconstruction can be regarded as significantly harder than just the reconstruction of a normal sinus rhythm.

9.4.2 Time Series Similarity Factors

ECG recordings are generally composed of more than a single lead, as in the 12-lead ECG, and a MECG which has 31 or 35 leads. In the terminology of time series analysis, the measurements from a single lead are called a univariate time series (UTS), while measurements from more than one lead are referred to as multivariate time series (MTS) [19].

For the purpose of defining measures of similarity between MECGs (between MTS), two MECG measurements are represented as:

$$\begin{aligned} X_1 &= \{X_1(1),\ldots,X_1(i),\ldots,X_1(m)\} \\ X_2 &= \{X_2(1),\ldots,X_2(i),\ldots,X_2(m)\}, \end{aligned} \tag{9.6}$$

where $X_1(i)$ and $X_2(i)$ are ith leads (i.e., UTS) of the first and second MECG measurements. m is the total number of leads, i.e., the length of the MTS.

The data from a lead are UTS, defined as a series of observations (measurements or data samples) made sequentially through time:

$$X(i) = \{X(i)_1,\ldots,X(i)_n\}, \tag{9.7}$$

where n is the total number of observations.

In order to evaluate the difference between reconstructed and target ECG, which are both MTS, we have used various similarity measures, described in the rest of this section.

The *Euclidean distance* between two MTS X_1 and X_2 of equal lengths can be defined as the mean distance of the corresponding UTSs [20]:

$$ED(X_1,X_2) = \frac{\sum_{k=1}^{m} d(X_1(k),X_2(k))}{m}, \tag{9.8}$$

where $d(X_1(k),X_2(k))$ is the Euclidean distance between the two UTS defined by:

$$d(X_1(k),X_2(k)) = \sqrt{\sum_{l=1}^{n} (X_1(k)_l - X_2(k)_l)^2}. \tag{9.9}$$

Correlation is a measure of statistical dependence of one MTS on another. Values of correlation near 0 often indicate that the variables are uncorrelated, while values near 1 or −1 indicate a strong positive or negative correlation.

The correlation between two MTS X_1 and X_2 can be defined as:

$$CORR(X_1,X_2) = \frac{\sum_{k=1}^{m} corr(X_1(k),X_2(k))}{m}, \tag{9.10}$$

where $corr(X_1(k),X_2(k))$ is the correlation between two UTS defined as [21]:

$$corr(X_1(k),X_2(k)) = \frac{\sum_{l=1}^{n} \left(\left(X_1(k)_l - \overline{X_1(k)}\right)\left(X_2(k)_l - \overline{X_2(k)}\right)\right)}{(n-1)s_{X_1(k)}s_{X_2(k)}}, \tag{9.11}$$

where $\left(X_1(k)_l - \overline{X_1(k)}\right)$ is the deviation of an individual UTS observation $X_1(k)_l$, from the UTS mean $\overline{X_1(k)}$, and $s_{X_1(k)}$ is the standard deviation of $X_1(k)$ UTS, and the same for $X_2(k)$.

The *PCA similarity factor* (S_{PCA}) is the measure of distance between the two MTS. It requires matrix representation of data, so that every column represents a single UTS from the MTS in question. To calculate S_{PCA} it is necessary to obtain the principal components (PCs) for each matrix, i.e., each MTS, and to choose the first *k* PCs using one of the heuristic approaches. For example, the PCs can be sorted by their variances and the first of them are selected whose sum of variances represent 95% of the total variance. S_{PCA} reflects the similarity of the first *k* PCs. The S_{PCA} of two MTS X_1 and X_2 is defined as follows [22]:

$$S_{PCA}(X_1, X_2) = \sum_{p=1}^{k} \sum_{q=1}^{k} \cos^2(\theta_{pq}), \tag{9.12}$$

where θ_{pq} is the angle between the pth PC of X_1 and q^{th} PC of X_2.

9.4.3 Principal Component Analysis

PCA is a multivariate statistical technique whose purpose is to condense the information of a large set of correlated variables into a few variables called principal components (PCs), while not losing the variability present in the original data set. The PCs are derived as a linear combination of the variables of the data set, with weights chosen so that the PCs become mutually uncorrelated. Each component contains new information about the data set, and is ordered so that the first few components account for most of the variability. In signal processing applications, PCA is performed on a set of time samples rather than on a data set of variables [23].

If we represent a MECG measurement by a matrix $X = [n \times m]$, where *n* is the number of time samples and *m* is the number of leads, then a covariance matrix is defined as:

$$C_X = \frac{1}{n-1} X_C^T X_C = \begin{bmatrix} c_{11} & \cdots & c_{1m} \\ \vdots & c_{ij} & \vdots \\ c_{m1} & \cdots & c_{mm} \end{bmatrix}, \tag{9.13}$$

where X_C is centralized MECG measurement X, obtained by subtracting the lead's mean from every lead, and $c_{ij} = c_{ji}$ is the covariance between the *i*th and *j*th leads.

The covariance matrix is diagonal in a coordinate system defined by its eigenvectors [24], so if the base transformation matrix is defined as:

$$T = \left[e_1 \ldots e_i \ldots e_m\right], \tag{9.14}$$

where e_i is ith eigenvector of C_X, then the covariance matrix in this base becomes:

$$C_X' = T^{-1} C_X T = \begin{bmatrix} \lambda_1 & 0 & \cdots & & 0 \\ 0 & \ddots & & & \\ \vdots & & \lambda_i & & \vdots \\ & & & \ddots & \\ 0 & & \cdots & & \lambda_m \end{bmatrix}, \tag{9.15}$$

where λ_i is the ith eigenvalue that corresponds to e_i. e_i are also called principal components of X [25].

Before presenting the original data in the new base, the columns of C_x' are sorted with respect to their variance, from the largest to the smallest variance. Sorted C_x' has $\lambda_i \geq \lambda_j$ if $j > t$ and is denoted by CS_X'. After sorting the eigenvalues, the related eigenvectors are also sorted. If the positions of the ith and jth eigenvalues in C_x' are changed, then the positions of the ith and jth eigenvectors in T have to be changed accordingly. Sorted T is denoted by TS. The original data matrix X is expressed in a new base defined by TS [26] as:

$$(X')^T = TS^{-1} \cdot X^T. \tag{9.16}$$

A new orthogonal coordinate system is chosen in which the covariance matrix is diagonal with no covariance between columns of X'. The existing variances are sorted from the largest to the smallest, with the largest variances in the first columns of X'. Systems with a large number of measured variables are often driven with just a few hidden variables. This results in the X' matrix having significant variances for only the first few columns. The first two PCs for input data obtained from a 31-lead MECG (described in Section 9.4.1), and input data projected on the coordinate system defined by the first two PCs, PC1 and PC2, are shown in Fig. 9.8.

By utilizing PCA, system variability can be analyzed by analyzing only the first few significant dimensions of a new base defined by PCs. Often, more than 95% of a heart's electrical activity can be explained by just the first three PCs (Fig. 9.9).

9.4.3.1 PCA-Based Reconstruction of 12-Lead ECG

For the purpose of reconstruction, we assume that two test MECG measurements are available for each person. The first MECG measurement is used for calculating PCs and selecting optimal unipolar leads. The second MECG measurement is used first as a data source for a real measurements from the determined optimal unipolar leads, and second, for the generation of the target 12-lead ECG that will be compared to the reconstructed ECG. We will denote these two MECG measurements as X_1 and X_2. The procedure for the reconstruction of a 12-lead ECG from two MECGs, by using PCA is described in the rest of this section.

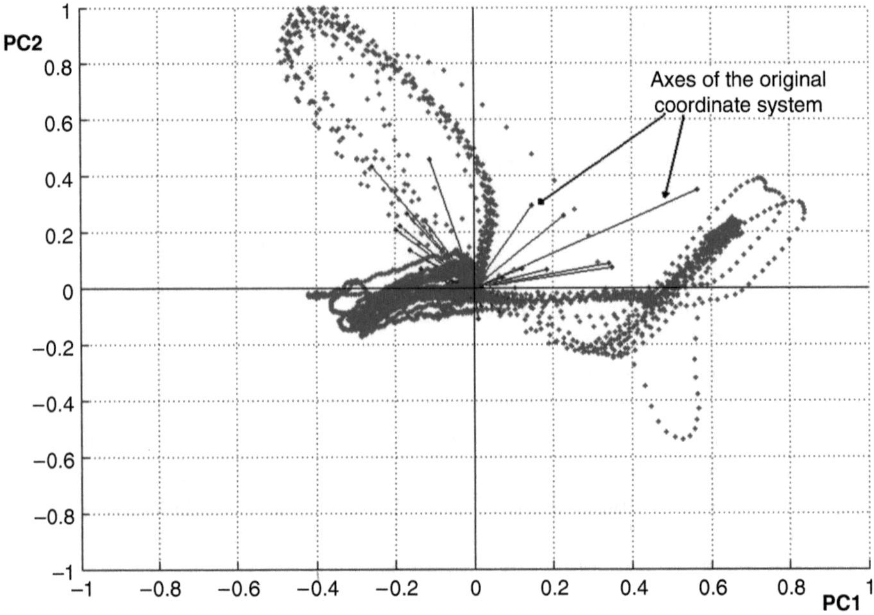

Fig. 9.8 Principal components (PCs) and input data in the base defined by the first two PCs. *Dots* are data points and *lines* are axes of the original coordinate system

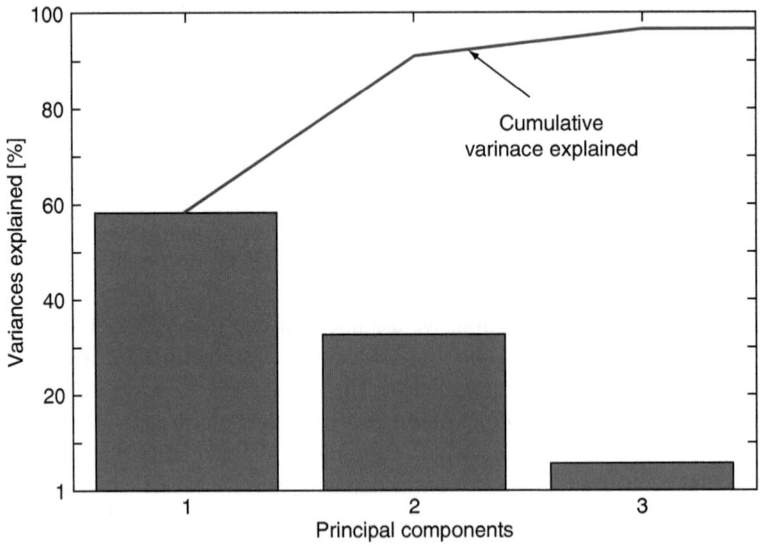

Fig. 9.9 Heart's electrical activity variance explained by the first three PCs

In the *first step* the matrix TS is generated from X_1 by the procedure described in Section 9.4.3. This matrix is denoted by TS_1.

Only the set of unipolar leads used for the reconstruction has to be taken into account. From now on they will be called reconstruction leads. X_2 is a set of leads:

$$X_2 = \{X_2(1), \ldots, X_2(i), \ldots, X_2(m)\}, \tag{9.17}$$

where m is the number of leads.

For the reconstruction method only a subset of $X_2$is used, e.g., for four reconstruction leads denoted as:

$$A = \{X_2(i), X_2(j), X_2(k), X_2(l)\}, \quad i, j, k, l \in \{1, \ldots, n \tag{9.18}$$

Although we concentrated on four reconstruction leads, their number is arbitrary but must be less than m. The selection procedure for reconstruction leads is described in Section 9.4.3.2.

Lead sets X and A can alternatively be given in a matrix form:

$$A = [n \times 4], X_2 = [n \times m], \tag{9.19}$$

where n is the number of samples. The indices of the reconstruction leads from the X_2 leads set form a set of indices:

$$I = \{i, j, k, l\}. \tag{9.20}$$

In the *next step* of the algorithm the information contained in A, is moved to the coordinate system defined by PCs. That is accomplished according to Eq. (9.16), which requires a matrix TS_2 (a sorted eigenvector matrix of X_2) obtained from the second MECG. Since X_2 would not exist if only the reconstruction leads were used, we cannot calculate TS_2. However, in our previous work [27], we have shown that the PCs may be considered constant for a person, if nothing drastic, such as heart surgery happens to the heart. We can formulate this statement in the form:

$$TS_1 = TS_2. \tag{9.21}$$

Equation (9.16) also expects X_2 as the last operand. Only reconstruction leads A of set $X_2$are measured, so we define a new matrix AE, with the same size as X_2, but with all lead vectors equal to 0, except the leads from the set A, i.e., the leads that have indices from the set I:

$$\begin{aligned} AE(j) &= X_2(j) \text{ } if \text{ } j \in I \\ AE(j) &= 0 \text{ } if \text{ } not(j \in I), \end{aligned} \tag{9.22}$$

where $AE(j)$ and $X_2(j)$ are the jth lead vector (jth columns) of AE and X_2, respectively.

By utilizing Eqs. (9.21) and (9.22), Eq. (9.16) becomes:

$$(AE')^T = (TS_1)^{-1} \cdot AE^T, \tag{9.23}$$

where AE is the AE matrix represented in the TS_1 coordinate system. A hypothesis of this algorithm is that AE' can be considered as an approximation of the unknown X_2(presented in the TS_1 coordinate system).

The *last step* of the algorithm will return AE' to the original coordinate system [26]:

$$(AE)^T = TS_1 \cdot (AE')^T, \tag{9.24}$$

where AE is an approximation of the X_2(measured MECG). The reconstructed 12-lead ECG can be extracted from AE. PCs calculated from the X_1 and data from only four optimal leads, obtained from X_2, were used to reconstruct 12-lead ECG. The target 12-lead ECG, extracted directly from X_2, is used to evaluate the obtained reconstructed results.

The described methodology is based on the assumption that the optimal leads are known. The procedure for their determination is described in the next section.

9.4.3.2 Optimal Leads Selection with PCA

The optimal leads selection algorithm, based on PCA, is shown in the flowchart (Fig. 9.10).

The trine participation of the MECG can be observed in the above algorithm. First, it is used to calculate PC; second, to generate the target 12-lead ECG; and third, to form all possible sets of four reconstruction leads. For every set of four reconstruction leads, a reconstructed 12-lead ECG is calculated according to the procedure described in Section 9.4.3.1. All reconstructed 12-lead ECGs are compared to the target 12-lead ECG, according to the similarity measures described in Section 9.4.2 and classified by the similarity factors. The four reconstruction leads that result in the best similarity factor are considered to be the best leads for the reconstruction.

Different similarity measures (see Section 9.4.2) have usually been observed to result in the same set of best leads [27]. It can be concluded that all the described similarity measures are acceptable; therefore, anyone of them can be selected.

9.4.4 Multivariate Linear Regression

An alternative approach, based on Multivariate Linear Regression (MLR), has been used to model the relationship between a 12-lead ECG and a set of reconstruction leads composed of three differential leads, each implemented by two closely placed electrodes, as described previously. In Section 9.4.3.1, the MECG was denoted as:

$$X = \{X(1), \ldots, X(i), \ldots, X(j), \ldots, X(m)\}. \tag{9.25}$$

As already explained, the differential lead (DL) is the difference between two unipolar leads:

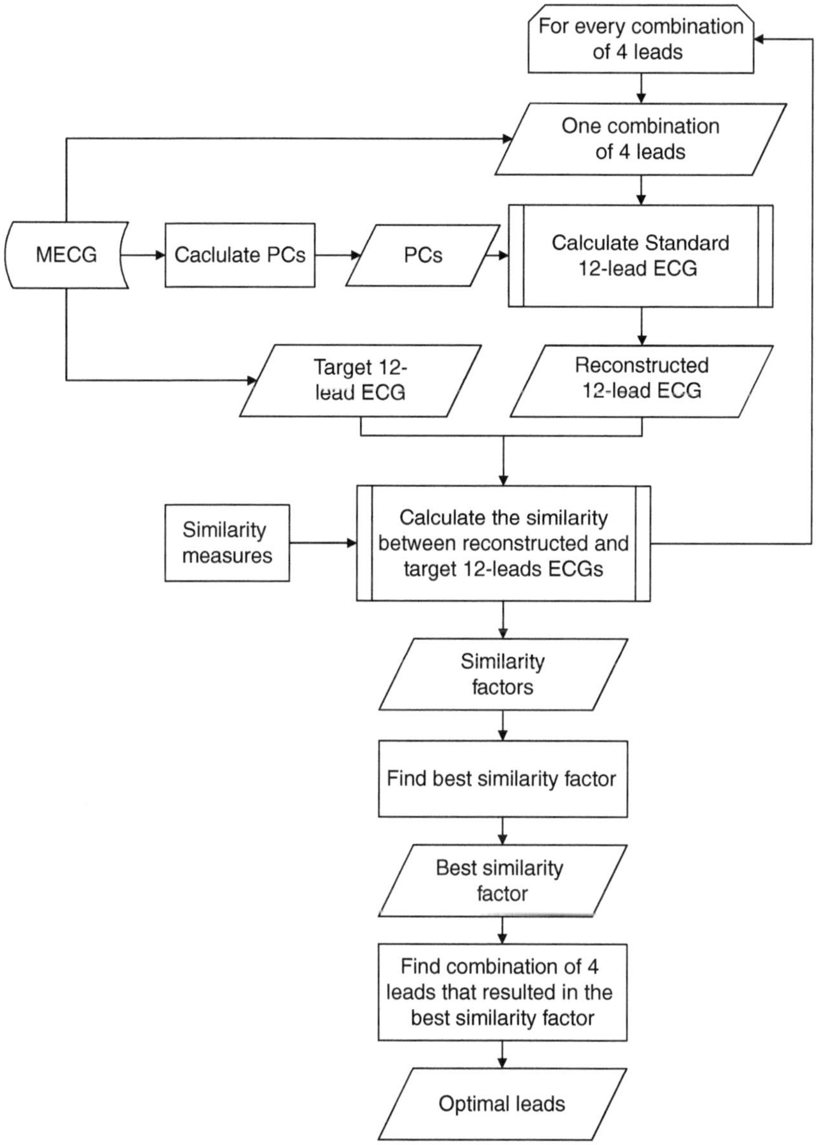

Fig. 9.10 Optimal leads selection with principal component analysis (PCA)

$$DL = X(i) - X(j). \tag{9.26}$$

For the MECG with 31 leads there are $\binom{31}{2} = 465$ possible differential leads. However, in WSN applications, the implementation of a small wireless electrode, whose measurement is emulated by DL, dictates the minimization of the distance between electrodes of a DL. That requirement restricts the set of all possible DLs to the set formed by a selected electrode and its nearest neighbors on the body surface. In the case of the MECG with 31 electrodes, this set contains just 81differential leads; however, there are $\binom{81}{3} = 85320$ possible combinations of three reconstruction leads denoted by B_{total}.

Let a set of three arbitrary bipolar differential leads be denoted as:

$$B = \{B(1), B(2), B(3)\}. \tag{9.27}$$

As explained in Section 9.2.1, 12-lead ECG can be represented as a set of 12-leads:

$$ECG12 = \{I, II, III, aVR, aVL, aVF, V1, V2, V3, V4, V5, V6\}. \tag{9.28}$$

As already noted, every MECG contains enough leads to generate the measured standard 12-lead ECG, so $ECG12$ is extracted from X (Eq. (9.25)) and represents a target ECG for reconstruction.

Generally, a linear regression model represents the relationship between a response (i.e., criterion variable) $ECG12$ and a predictor B [28]:

$$ECG12 = \alpha_1 f_1(B) + \ldots + \alpha_j f_j(B) + \ldots + \alpha_p f_p(B) + \varepsilon \tag{9.29}$$

The response is modeled as a linear combination of functions (themselves not necessarily linear) of the predictor, plus a random error ε. The expressions $f_j(B), (j = 1,\ldots,p)$ are the terms of the model while $\alpha_j, (j = 1,\ldots,p)$ are the coefficients. Errors ε are assumed to be uncorrelated and distributed with mean 0, and constant, but unknown, variance. Our problem can be covered by the multivariate regression due to the fact that the response variable $ECG12$ is multidimensional, i.e., it is composed of 12 leads (variables) [29].

Given n independent observations (samples), $(B_1, ECG12_1), \ldots, (B_n, ECG12_n)$, of the predictor B and the response $ECG12$, the linear regression model becomes an n-by-p system of equations:

$$\begin{bmatrix} ECG12_1 \\ \vdots \\ ECG12_n \end{bmatrix} = \begin{bmatrix} f_1(B_1) & \cdots & f_p(B_1) \\ \vdots & & \vdots \\ f_1(B_n) & \cdots & f_p(B_n) \end{bmatrix} \cdot \begin{bmatrix} \alpha_1 \\ \vdots \\ \alpha_p \end{bmatrix} + \begin{bmatrix} \varepsilon_1 \\ \vdots \\ \varepsilon_n \end{bmatrix}, \tag{9.30}$$

or

$$ECG12 = M \cdot \alpha + \varepsilon, \tag{9.31}$$

where M is the design matrix of the system. The columns of M are the terms of the model evaluated at the predictors. To fit the model to an input data, the above system must be solved for the p coefficient vectors: $\begin{bmatrix} \alpha_1 & \ldots & \alpha_p \end{bmatrix}^T$. We solved the system by applying the least-squares solution separately for every lead of $ECG12$ and therefore reducing the multivariate regression model to 12 multiple regressions [30], applying the MATLAB *regress* function [29].

The predictor B is multidimensional because it is composed of 3 variables, as are the functions f_j that form the terms of the model. For a three-dimensional predictor $B = \{B(1), B(2), B(3)\}$, terms for the model could include $f_1(B) = B(1)$ (or for example, $f_1(B) = B(2)$) (linear term), $f_2(B) = B(1)^2$(quadratic term), and $f_3(B) = B(1) \bullet B(2)$ (pairwise interaction term). Typically, the function $f(B) = 1$ is included among f_j, so that the design matrix M contains a column of ones and the model contains a constant, y-intercept, term.

We have explored the use of linear additive models. They have terms $f(B) = 1$ and $f(B) = B(i)$. In the case of three bipolar leads, the design matrix M of the linear additive model is:

$$M = \begin{bmatrix} 1 & B(1)_1 & B(2)_1 & B(3)_1 \\ \vdots & \vdots & \vdots & \vdots \\ 1 & B(1)_n & B(2)_n & B(3)_n \end{bmatrix}, \tag{9.32}$$

with four coefficients in the vector $\alpha = [\alpha_1 \quad \alpha_2 \quad \alpha_3 \quad \alpha_4]$ that are obtained after the system solution. If we denote the solution coefficients by α_{res}, then the result of the $M \bullet \alpha_{res}$ is the best approximation of $ECG12$ in the sense of the least-square solution, obtained from the three selected differential leads B.

For reconstruction purposes, only the best combination of differential leads is required. Section 9.4.4.1 explains how such a combination of three differential leads is selected from the set B_{total}. If the system (9.31) is solved for the three optimal differential leads, then $M \bullet \alpha_{res}$ is considered to be the best approximation to the target 12-lead ECG. If the wireless electrodes, which implement differential leads, are positioned on the same places on the body surface as optimal differential leads, the calculated coefficients vectors can be applied on the measurements from wireless electrode to reconstruct a 12-lead ECG.

9.4.4.1 Optimal Differential Leads Selection for MLR

In the previous section the set of all possible combinations of three differential leads B_{total} was introduced. In order to find an optimal combination of three differential leads, a "brute force attack" type of algorithm was used: MLR is conducted for every element of the set B_{total}. The optimal combination of three differential leads is considered to be the one that reconstructs a 12-lead ECG that has the strongest correlation with the target 12-lead ECG. This procedure is illustrated in Fig. 9.11.

9.4.5 Personalization

The optimal lead set and the transformation matrix for reconstructing the 12-lead ECG are personalized, in the sense that a different optimal lead set is found for every patient and a different transformation matrix is calculated for every patient.

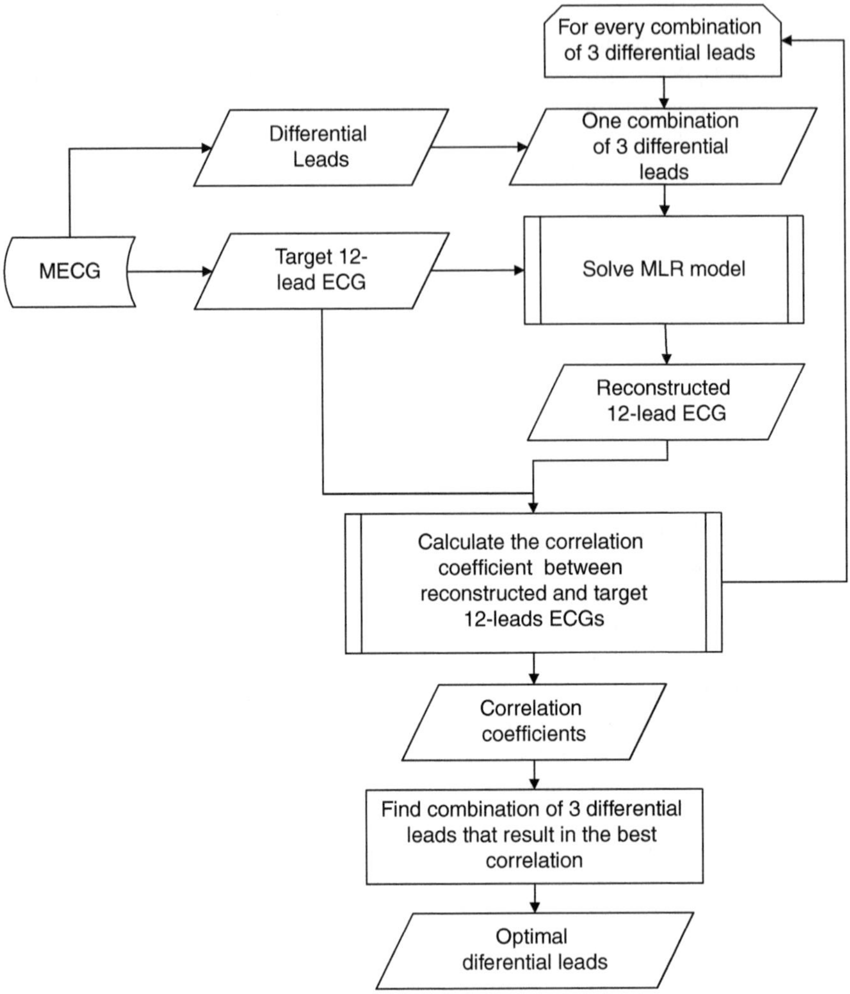

Fig. 9.11 Optimal leads selection with Multivariate Linear Regression (MLR)

In the case of PCA, the transformation matrix is the matrix defined by PCs, and in the case of MLR the transformation matrix is composed of the coefficient vectors.

By studying a sufficiently large number of cases for different patients it should be possible to find optimal leads that will be suitable for most patients and to calculate a global transformation that gives, on average, the best possible fit for each individual case [31]. Although possible, such an approach is not necessary in WSN applications because wireless electrodes have a lot of positional freedom, since they are not limited by the connection wires of an ECG device.

9.5 Validation of the Proposed Algorithms

9.5.1 PCA-Based Approach

The proposed PCA-based method for selecting optimal reconstruction leads and for reconstructing 12–lead ECG will be illustrated and validated on the real MECG measurement. The data used are described in Section 9.4.1. By applying the algorithm described in Section 9.4.3.2, we obtained, for a particular person, with no additional restrictions, the optimal reconstruction lead set: {6, 8, 13, 15}. Their positions in the 31-channel MECG placing are denoted in Fig. 9.12 by squares. The corresponding reconstructed 12-lead ECG is given in Fig. 9.13 in red and the target ECG in black.

9.5.2 Linear-Regression-Based Approach

The MLR method for selecting the optimal leads and for reconstructing the 12–lead ECG has also been validated. The data used are described in Section 9.4.1. The optimal differential lead set composed of three differential leads and obtained by the methodology described in Section 9.4.4 is: {(9,13), (15,19), (20,22)}. Positions of these optimal leads are denoted in Fig. 9.14 by lines.

The MLR-based algorithm is seen to select optimal differential leads, so that they "view" the heart from all three spatial dimensions (Fig. 9.14). The corresponding reconstructed 12-lead ECG is given in Fig. 9.15, again with the target ECG.

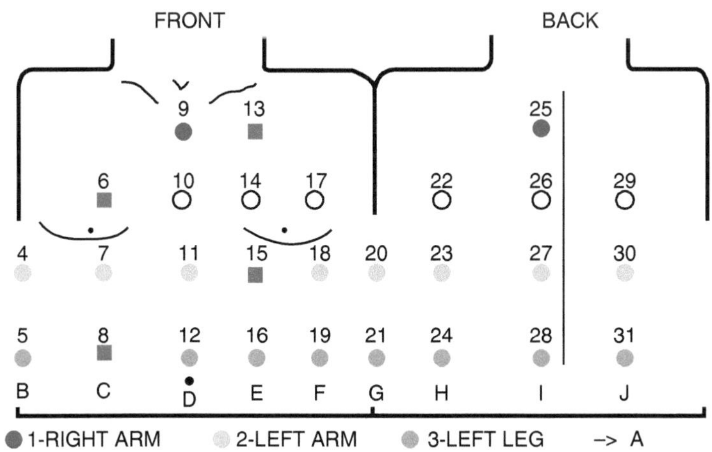

Fig. 9.12 Unipolar electrode placing for principal component analysis (PCA) based reconstruction of the 12-lead electrocardiogram (ECG). Optimal unipolar leads {6,8,13,15} are marked by *squares*

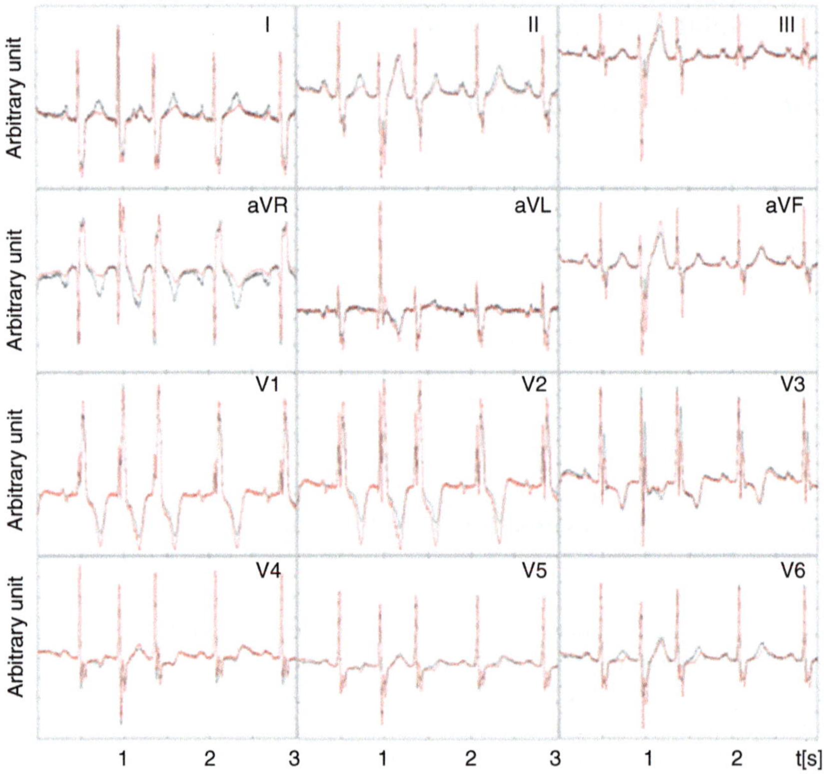

Fig. 9.13 Target (*black*) and principal component analysis (PCA) reconstructed (*red*) 12-lead electrocardiograms (ECGs)

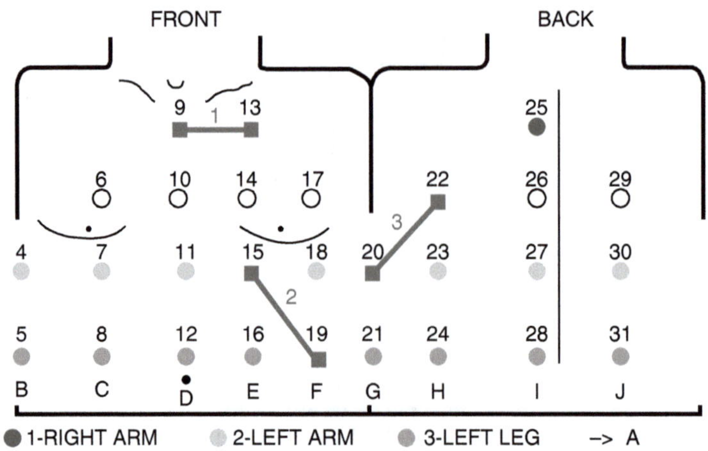

Fig. 9.14 Differential lead (DL) positions used by the Multivariate Linear Regression (MLR) based reconstruction of the 12-lead electrocardiogram (ECG). The *squares* connected with *lines* denote differential leads (DLs) 1, 2, and 3

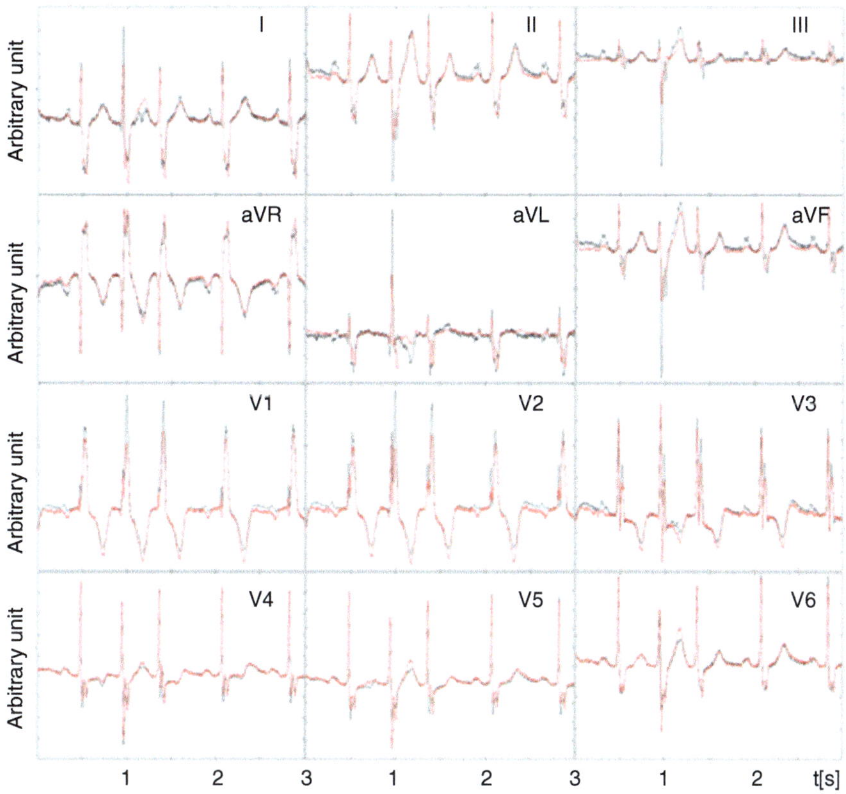

Fig. 9.15 Target (*black*) and Multivariate Linear Regression (MLR) reconstructed (*red*) 12-lead electrocardiograms (ECGs)

9.6 Conclusion

We have shown that the proposed methodology for measuring the electrical activity of the heart enables improved patient mobility and comfort, requiring only a minimal set of small wireless electrodes. Two different approaches to the 12-lead ECG reconstruction from a limited number of leads have been demonstrated, PCA based on unipolar leads, and MLR based on differential leads. High-resolution multichannel ECG measurements were used as the input data. From the set of all possible differential leads only those with closely placed electrodes were taken into account, since they enable a simple implementation of wireless electrodes that are connected in a WSN.

Based on the two described statistical techniques, together with the use of different time series similarity factors, optimal leads and a corresponding transformation matrix have been selected for every patient. The PCA-based methodology selects four best unipolar reconstruction leads, which makes it applicable for systems with reduced number of wired electrodes, connected to a gateway that transmits the measured signals wirelessly, for example, to a personal terminal.

The MLR methodology, however, selects three best differential leads that can be measured with wireless electrodes and transmitted in a wireless method to a personal terminal. In the case of PCA, the transformation matrix is defined by PCs, while in the case of MLR the transformation matrix is composed of the coefficient vectors. The transformation matrix is applied on the measurements from the leads selected by the algorithm to reconstruct the 12-lead ECG.

Both approaches were verified on the same input MECG that contained a single supraventricular extrasystole. It can be concluded that the reconstructed and target 12-lead ECGs show good agreement throughout all 12-leads. We have confirmed with several additional verifications of normal ECGs in sinus rhythm that the proposed PCA and MLR methodologies provide even better agreement with the target 12-lead ECGs.

We plan to verify and evaluate the proposed approaches on even more test cases, in order to investigate their diagnostic ability in more details. The wireless approach, supported by MLR, has an important future in the field of patient monitoring, due to the fact that patient mobility and comfort are increased. Hopefully, our work will contribute to the penetration of the WSN technology into the field of the electrocardiography.

References

1. Finlay DD et al (2007) Synthesising the 12-lead electrocardiogram: trends and challenges. Eur J Intern Med 18:566–570
2. Dower GE, Yakush Andrew, Nazzal SB, Jutzy RV, Ruiz CE (1988) Deriving the 12-lead electrocardiogram from four (EASI) electrodes. J Electrocardiol 21(1):182–187
3. Wei D (2002) Deriving the 12-lead electrocardiogram from four standard leads using information redundancy in the 12-lead system. J Bioelectromagn 2(4):127–135
4. Drew BJ, Pelter MM, Brodnick DE, Yadav AV, Dempel D, Adams MG (2002) Comparison of a new reduced lead set ECG With the standard ECG for diagnosing cardiac arrhythmias and myocardial ischemia. J Electrocardiol 35:S13–S21
5. Nelwan SP, Kors JA, Meij SH (2000) Minimal lead sets for reconstruction of 12-lead electrocardiograms. J Electrocardiol 33:S163–S166
6. Roman Trobec (2003) Computer analysis of multichannel ECG. Comput Biol Med 33(3):215–226
7. Trobec R, Depolli M, Avbelj V (2010) Wireless network of bipolar body electrodes. In: Proceeding of 7th international conference on wireless on-demand network systems and services, WONS, Kranjska Gora, pp 145–150
8. Malmivuo J, Plonsey R (1995) 12-Lead ECG system, vol Bioelectromagnetism – principles and applications of bioelectric and biomagnetic fields. Oxford University Press, New York, ch. 15
9. Ilyas M, Imad M (eds) (2005) Energy management. In: Handbook of sensor networks: compact wireless and wired sensing systems. CRC Press, Boca Raton, FL, ch. 6
10. Merja Puurtinen, Jari Viik, Jari Hyttinen (2009) Best electrode locations for a small bipolar ECG device: signal strength analysis of clinical data. Ann Biomed Eng 37(2):31–36
11. Yang G-Z (ed) (2006) Body sensor networks. Springer-Verlag, London
12. Wood A et al (2008) Context-aware wireless sensor networks for assisted living and residential monitoring. IEEE Network 22(4):26–33
13. Breamar (2006) DL800 Operator Manual http;//www.braemarinc.com/pdfs/dl800manual.pdf

14. Nelwan SP, Carey MG (2008) Reduced and alternative lead sets: clinical implications and technical challenges. J Electrocardiol 41(6):451–3
15. Bu G, Berbari JE (2006) A single moving dipole model of ventricular depolarization. Biomed Sci Instrum 42:237–242
16. Nelwan SP (2004) Reconstruction of the 12-lead electrocardiogram from reduced lead sets, Journal of Electrocardiology, 37:11–18
17. Nelwan SP et al (2004) Simultaneous comparison of three derived 12-lead ECGs with standard ECG at rest and during percutaneous coronary occlusion. J Electrocardiol 37:S171–S172
18. Robertson AJ, Clark E, van Oosterom A, MAcFarlane PW (2002) ECG monitoring using a limited lead set. J Bioelectromagn 4:295–299
19. Brockwell PJ, Davis RA (1991) Multivariate time series. Time series: theory and methods, 2nd edn. Springer, New York, ch. 11
20. Lee S, Chun S, Kim S, Lee J, Chung C (2000) Similarity search for multidimensional data sequences. In: Proceedings of the 16th international conference on data engineering, San Diego, CA, p 599
21. Kachigan SK (1991) Correlation analysis. Multivariate statistical analysis (a conceptual introduction), 2nd edn. Radius Press, New York, pp 130–133, ch. 3
22. Yang K, Shahabi C (2004) A PCA-based similarity measure for multivariate time series. In: Proceedings of the 2nd ACM international workshop on multimedia database, Arlington, VA, pp 65–74
23. Castells F, Laguna P, Sörnmo L, Bollmann A, Roig JM (2007) Principal component analysis in ECG signal processing. EURASIP J Adv Signal Process 2007:1–22
24. Hoffman K, Kunze R (1971) Normal operators. Linear Algebra, 2nd edn. Prentice Hall, Englewood Cliffs, NJ, ch. 8.5
25. Bilodeau M, Brenner D (1999) Principal components. Theory of multivariate statistics. Springer, New York, ch. 10
26. Hoffman K, Kunze R (1971) Coordinates. Linear algebra, 2nd edn. Prentice Hall, Englewood Cliffs, NJ, pp 49–54
27. Tomasic I, Skala K, Trobec R (2008) Principal component analysis and visualization in optimization and personalization of lead's set for generation of standard 12-lead ECG. In: Proceedings of the MIPRO international convention, Vol. 1., Opatija, pp 307–313
28. Draper NR, Smith H (1998) The general regression situation. Applied regression analysis, 3rd edn. Wiley-Interscience, New York, ch. 5
29. Inc. The MathWorks (2009) Statistical toolbox: linear regression. MATLAB documentation
30. Bilodeau M, Brenner D (1999) Multivariate regression. Theory of multivariate statistics. Springer, New York, ch. 9
31. Horacek BM, Warren JW, Field DQ, Feldman CL (2002) Statistical and deterministic approaches to designing transformations of electrocardiographic leads. Journal of electrocardiology, 35:41–52

Chapter 10
Vehicular Sensor Networks: General Aspects and Implementation Issues

Vladimir Atanasovski and Liljana Gavrilovska

10.1 Introduction

The usage of sensors and sensor networks to sense road and vehicle phenomena and send sensor data to relevant entities (e.g., users, institutions etc.) gives rise to the concept of *Vehicular Sensor Networks (VSNs)*. VSNs are a subset of Vehicular Ad-hoc NETworks (VANETs) placed either in vehicles or alongside roads creating an end-to-end reliable network for disseminating sensor data gathered from a vehicular environment. Depending on the communication path, VSNs encompass Vehicle-to-Infrastructure or Infrastructure-to-Vehicle (V2I and I2V, respectively) architectures and Vehicle-to-Vehicle (V2V) architecture. The former deals with communication between sensors from a vehicular environment and specific Road Side Units (RSUs) or external gateways, whereas the latter one defines the communication among sensors in a vehicular environment in a pure ad hoc, i.e., infrastructureless, fashion. The VSNs have been widely investigated and proved to be very useful for supporting car and road safety, traffic pattern analysis, road surface diagnosis, urban environmental monitoring, street-level air pollution monitoring, and many other transportation application systems. Their increasing popularity emphasizes the need for thorough analysis of their most relevant features in order to pave the way toward VSNs' full practical deployment.

VSNs can be, generally speaking, also regarded as Wireless Sensor Networks (WSNs) with some notable differences [1]. First and foremost, power consumption is not a critical issue due to the vehicles' battery that can produce large enough amounts of energy at a relatively low cost. However, a specific problem that arises due to the vehicular environment is the VSNs' *dynamic topology*. The dynamic topology, caused by the mobility of nodes, creates a new set of challenges that need to be resolved. Furthermore, a fully functional VSN needs to work properly under various traffic conditions (i.e., high and low traffic density)

V. Atanasovski (✉) and L. Gavrilovska
Faculty of Electrical Engineering and Information Technologies, Skopje
e-mail: vladimir@feit.ukim.edu.mk; liljana@feit.ukim.edu.mk

L. Gavrilovska et al. (eds.), *Application and Multidisciplinary Aspects of Wireless Sensor Networks*, Computer Communications and Networks,
DOI 10.1007/978-1-84996-510-1_10,

making the network *scalability* another challenging problem. It is usually met by using mesh networking (in high traffic density situations) and long ranged transmissions (in low-density situations). In addition to scalability, *routing,* and maintaining *end-to-end connectivity* are additional requirements of a functional VSN, the latter being especially important for sparsely populated VSNs where vehicles form isolated clusters as a result of the dynamic network topology. Possible solutions include the use of specific roadside gateways used for proper data propagation and message relay boxes for storing messages [2]. Since the VSNs' main applications are safety applications, *reliability* and *promptness* are crucial to the feasibility of these applications. This is a serious challenge that needs to be resolved for every specific application. Last but not least, the inclusion of *Quality-of-Service* (*QoS*) in VSNs is challenging, considering the high mobility of the VSN nodes.

The importance and the rising popularity of VSNs lead to numerous international research projects lately. The main interests of the majority of VSNs-related projects lie in the design of a cooperative vehicular environment enabling a plethora of cooperative services for its participants and design of efficient addressing schemes. For example, the CVIS project [3] aims to create a unified technical solution for cooperative VSNs that addresses issues such as user acceptance, data privacy and security, system openness and interoperability, risk and liability, public policy needs, business models etc. The SafeSpot project [4] targets an integration among vehicle-isolated, telematic, technologies into a single cooperative solution that enables development of reliable and extended driving support systems for road safety. Both projects also focus on dynamic creation of road maps, a feature that may be extensively used for intelligent transportation applications. The Mobile Millenium project [5] focuses on the reliability aspect of VSNs and exploits a centralized packet delivery solution for supporting safety-critical applications. The GeoNet project [6] implements and formally tests a networking mechanism as a stand-alone software module for cooperative systems. It relies on geographical addressing scheme that enables a multi-hop communication among vehicles and the infrastructure. This project fills the implementation gap among many finished and ongoing VSNs-related projects. Its prominent work inspired the latest spatiotemporal addressing [7], a natural extension of the geographical addressing.

This chapter elaborates the general aspects and the system implementation issues of VSNs. It provides a VSN classification framework, briefly discusses the basic mechanisms deployed in VSNs, introduces a layering architecture for VSNs, discusses the practical implementation of sensors in vehicular environments, introduces the RFID technology as an evolutionary step in VSNs' design and classifies and elaborates on the most popular applications of VSNs today. The chapter's aim is to serve as an introductory text in the field, tackling the specifics of the vehicular environment that distinguishes the VSNs within the wireless sensor networking paradigm in general. More details on this vibrant research area and the relevant entities that drive the VSNs' related research can be found in [8, 9].

10.2 VSNs Classifications

The numerous VSNs' types and specific implementations found in the literature today can be broadly classified into two groups:

- *Intra-vehicle VSNs* (a "single vehicle" VSN used for giving diagnostics to the driver)
- *Inter-vehicle VSNs* (used for sharing the sensor data with other vehicles or infrastructure components)

Intra-vehicle VSNs are introduced for the purposes of monitoring, control, and communication between components and subsystems inside a vehicle. They are motivated by the increasing complexity, weight, and cost of the wiring harness in addition to the versatility of the wireless networking, which opens room for novel architectures and reprogrammable functionalities. In-vehicle connectivity in an intra-vehicle VSN is usually enabled by wireless standards for ad hoc communications such as Bluetooth [10], ZigBee [11] and UWB [12].

Inter-vehicle VSNs, either V2I/I2V or V2V, are based on different DSRC (Dedicated Short Range Communication) standards at physical and data link layer [13]. They can either be fully distributed or centralized. In V2V communication, a vehicle can communicate with its neighboring vehicles even in the absence of a central entity (e.g., a Base Station). The concept of this direct communication is to send vehicle safety messages one-to-one or one-to-many via a wireless connection. Such messages are usually short in length and have very short lifetime in which they must reach the destination. The inter-vehicle communication system is gaining momentum lately contributing to the increase in safety and the more intelligent traffic management on roads [14].

Both types of VSNs have their specifics stemming from the actual technology in use and applications in focus. The intra-vehicle VSNs are limited in space (one vehicle) and applications (vehicle diagnostics such as road conditions, tire pressure, etc.) making them predictable and easier for implementation. As a result, there are numerous proprietary solutions in this field [8]. On the other hand, the inter-vehicle VSNs prove to be more challenging since they require many mechanisms for proper operation of the sensor data sharing under vehicular conditions. Some of the basic mechanisms needed for their comprehensive analysis are elaborated in the following section.

10.3 VSNs Basic Mechanisms

This subsection elaborates the basic mechanisms needed to fully comprehend some of the most prominent specifics of the VSNs. They comprise *topology formation*, *mobility*, *routing*, and *data dissemination*. Further aspects can be found in [8].

10.3.1 VSNs Topology

Creating high-speed, highly scalable, and secure VSNs presents an extraordinary challenge. The combination of highly dynamic mobility patterns in VSNs, combined with high velocities, results in highly dynamic network topologies. However, VSNs also employ certain features that can actually facilitate the topology formation process. They have access to ample computational and power resources within the network itself and can usually utilize high-performance wireless communication and advanced antenna technology. In addition, it can be expected that a significant fraction of vehicles will have an accurate knowledge of their own geographical position by means of GPS.

Main topology candidate for VSNs is the *Wireless Mesh Networking* (*WMN*) [15], which lately emerged as an important technology in building next generation networks. It provides a range of benefits over traditional wired and wireless networking solutions including low deployment costs, scalability under higher traffic density situations, and resiliency to faults. Moreover, the WMN is often described as being autonomic with self-healing and self-configuring properties and its popularity has grown both as a research platform and as a commercially exploitable technology. Therefore, WMN is today used as the key topology formation concept in VSNs, Fig. 10.1.

A WMN topology and the *multi-hop communication* it inherently provides offer multiple redundant communications paths throughout the network. This characteristic can be useful for overcoming dead-spots if some sensors in some vehicles stop operating or the communication is temporarily blocked, e.g., sudden lane change of a heavy vehicle (such as a truck or a bus) can introduce a shade zone and,

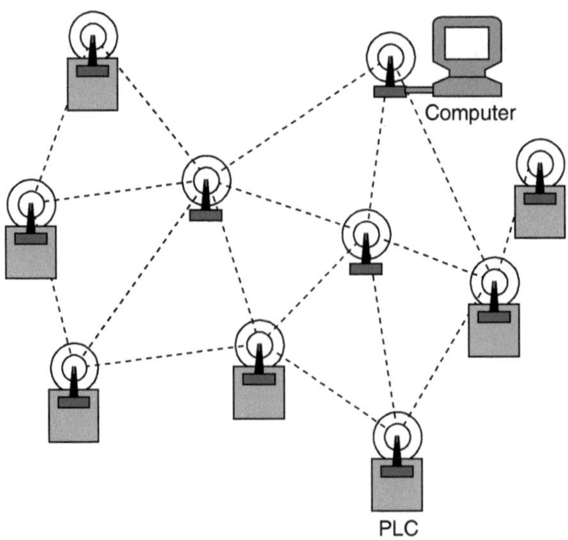

Fig. 10.1 WMN topology

consequently, loss of signal in the communication between fast vehicles. Furthermore, the WMN topology comprises shorter distances among network participants allowing for higher quality of the received signal. This makes the links more reliable without increasing transmitter power in individual nodes. Therefore, the mesh networking leads to increased network range, network redundancy, and improvement of the general network reliability, especially under increasing traffic density situations. Additional advantages of the WMN topology are in its self-configuration and self-healing capabilities. A WMN network does not need a system administrator or a manual configuration, so adding new nodes is simple. The network discovers the new node and automatically incorporates it into the existing system. Also, the control of the wireless system is *distributed* throughout the network, which allows properly equipped vehicles to communicate directly with other points of the network without being routed through a central control point.

The unique combination of multi-hop communication, WMN topology, and distributed control is the key for a whole new dimension of sensor communication. It is being completely exploited in VSNs.

10.3.2 Mobility in VSNs

VSNs are characterized by high mobility of nodes making the mobility issues an important aspect that should be carefully scrutinized when evaluating any VSN protocol. The VSNs' mobility models should involve road maps with all the constraints related to the vehicular environment in order to get accurate and realistic results. This section reviews the models that were proposed in the general MANET context, but are usable in VANETs and VSNs as well. More comprehensive study of the topic can be found in [16].

Different mobility models for VSNs differ in many parameters. Some of them use traffic control mechanisms at intersections, whereas others assume continuous movements at these points. Some protocols assume roads to be single lanes, while others support multilane roads. Additionally, some protocols define *security distance*, i.e., the minimal distance between vehicles that guarantees safety. This subsection briefly describes the characteristics of some of the most relevant mobility models applicable for VSNs.

Freeway is a map-based model where the simulation area represented by a generated map includes many freeways and many lanes in each freeway (there are no urban roads and intersections) [17]. The nodes, which are randomly placed on the lanes, move using history-based speeds following random acceleration. The distance that should be maintained between two subsequent vehicles in a lane is also defined. The model is unrealistic since the vehicles are not allowed to change the lanes and are forced to move on lanes until reaching the simulation area limit.

Manhattan is also generated-map-based model (containing vertical and horizontal roads with two lanes for moving in each direction) to simulate an urban environment. The vehicles are randomly deployed in the area and are moving with history-based speeds. When reaching the crossroads, the vehicles randomly choose a new direction, and, contrary to the previous model, the vehicles can change the lane at crossroads. The security distance is also used, but there is no control mechanism at crossroads where nodes continue to move without stopping making the model unrealistic [17].

City Section Model (CSM) is generated-map-based algorithm, where the nodes pause time and destination selection are randomly chosen. The speed of the nodes is constrained by the security distance, along with the maximum speed limit of the road. Similar features are applied to the Rice University Model [18], which differs in using real maps obtained from the TIGER/Lines database [18].

Stop Sign Model (SSM), contrary to the previous models, integrates a traffic control mechanism, where in every crossroad there is a stop signal that forces the vehicles to slow down and pause. SSM is based on real maps of the TIGER/Lines database, and all roads are assigned a single lane in each direction. Overtaking a vehicle is not allowed, and the vehicle should tune its speed to keep the security distance. The problem with this model is the unrealistic disposition of the stop signals since it is not realistic to expect to find stop signals at each intersection.

Traffic Sign Model (TSM) replaces the stop signals by traffic lights. When the first vehicle reaches the intersection, the light is randomly turned red with some probability p. Then, the vehicle pauses for a random time (pause time). After the delay, the light turns green and the nodes traverse the crossroads until the queue is empty. When the next vehicle arrives, the process is repeated. TSM is more stable than SSM since the pause time for stop signals is shorter than the one for traffic lights [19].

STRAW is also a model using real maps of TIGER/Line database and like the other models (except freeway), the roads include one lane in each direction and are divided into segments. The nodes are placed randomly at the beginning of the simulation, then they are moving using the car following model [20] and are accelerating until reaching the maximum speed of the segment. The overtaking is not allowed, but the security distance is maintained. At crossroads, the vehicles always slow down, even when they change a segment and they turn without a full stop, which is realistic. The traffic control mechanism defines the stop signs and traffic lights, and the routes are selected either randomly or toward a path that uses the shortest lane.

MOVE [21] is a tool built on top of an open source micro-traffic simulator named SUMO [22]. It incorporates GUIs to facilitate the process of road topology and vehicular mobility definition. The output of MOVE is a mobility trace file that contains the information of vehicle movements, which can be used by various simulation tools. MOVE consists of Map Editor and Vehicle Movement Editor. The first is used to create the road map, which can be created manually by users, automatically generated, or imported from existing real world maps. The second allows users to specify the trips of vehicles and the route that each vehicle will take for the trip.

A novel and realistic model is proposed by Gorgorin et al. in [23], where an overtaking mechanism is applied within multilane segments. Moreover, the model allows specifying the driver type, which affects many parameters of the vehicle, like speed and acceleration. The model includes traffic lights and stop signs at the intersections.

These mobility models are only a fraction of the possible VSN applicable ones. Being a vibrant research area, the quest for more realistic VSN mobility model continues.

10.3.3 Routing in VSNs

As it has been previously discussed, VSNs are ad hoc networks in a vehicular environment used for transmission of "sensed" data. Therefore, routing, which is a major issue in ad hoc networking in general, is also a serious challenge in VSNs as well. Many routing protocols and strategies for ad hoc networks have been proposed in recent years, including those specific for VANETs and VSNs. These protocols can be used for sensor data with the additional aspects on QoS due to the nature of applications that are mostly associated with VSNs.

In general, VSN routing techniques can be broadly distinguished into two main classes:

- *Topology based* [24]
- *Geo-position based* [25]

The first step in topology-based routing is recognizing the topology of the network. Then, distance vector and link state algorithms can be used for the maintenance of the routing tables in each node. Additionally, topology-based routing protocols may be divided into *proactive* and *reactive* ones. Proactive topology-based routing protocols periodically distribute their routing tables throughout the network so that the changes in the topology can be updated continuously, while reactive ones are event driven and routes are discovered only when they are needed. Proactive routing produces smaller delays for propagating the data through the network due to the already established routes. However, it uses large amounts of data for routing table maintenance and reacts slowly on topology restructurings or failures. Good example for proactive topology-based routing is FSR (Fisheye State Routing). On the other hand, reactive topology-based routing uses less data for routing table maintenance, which makes it more suitable for highly dynamic topologies such as the VSNs' type of scenarios. But, this type of routing can lead to higher latency and network congestions since it floods the network during route discovery. Reactive topology-based routing is represented by AODV (Ad-hoc On demand Distance Vector) and DSR (Dynamic Source Routing). Considering all the pros and cons of both topology-based routing protocols, it is evident that they do not meet all the demands that the VSNs may impose.

Geo-position-based routing is a better choice in a vehicular environment according to a number of studies. This routing technique is based on sending the information to physical locations instead of logical addresses. Each node possesses a unique identifier, which contains information of the geographical position of the node. The geographical location can be obtained either by GPS or specific distance sensors. GPS devices are expensive and impractical to use, which leads to the idea of acquiring the location of nodes through appropriate sensors. An intermediate node in geo-positional routing only needs to be aware of its own location and the geographical location of its one-hop neighbors, while the source node additionally needs to be aware of the destinations' geographic location.

Three main strategies in geo-positional routing can be distinguished, i.e., *flooding*, *single path,* and *multipath* [26]. When a flooding strategy is used, the network is being flooded starting from the source to its one-hop neighbors. When using a multipath strategy, multiple paths from source to destination are used in order to forward the information. When using a single-path strategy, a single path from the source to the destination is being established. The single-path strategy can be implemented in two ways, i.e., *greedy forwarding* and *face forwarding*. Greedy forwarding sends the information to the next nearest one-hop neighbor of the previous node or the next nearest one-hop neighbor in the direction to the destination node (*compass forwarding*). The face forwarding strategy recognizes the topology and by measuring the angles of the nearest one-hop neighbors, compared to the direction of the destination, sends the information through the interior of the network.

The advantages of geo-positional routing, compared to topology-based routing, are in the no route set up and route maintenance need, which leads to a reduction in latency. On the other hand, position-based routing requires complex data dissemination and harvesting mechanisms for successful management of position information so that communicating nodes can learn their respective locations.

10.3.4 Basic Dissemination Mechanisms in VSNs

Data dissemination in VSNs refers to a way of distributing the data collected from the vehicles' sensors among the vehicles and to the RSUs. It resides on top of the routing mechanisms and uses them to enable efficient transportation of data to the intended recipients while satisfying certain requirements such as delay, reliability, etc. These requirements vary depending upon the data being disseminated (safety or user applications).

Data dissemination in VSNs is a complex subject that is linked to the MAC and routing issues and brings additional possibilities such as the use of the infrastructure support and the possibility of aggregating the data. Different approaches of data dissemination in VSNs are found in the literature. Each of them has its own advantages and disadvantages. Depending on the architecture of the VSN (whether it is a V2I/I2V or V2V communication), four basic ways of data dissemination are known. For V2I/I2V, there are the *push-based* and *pull-based* mechanisms, whereas for V2V, there are the *flooding* and *relaying* (next hop) mechanisms.

The idea of *push-based* dissemination is useful for distributing so called popular data like traffic alerts, weather alerts, etc. The vehicle that wants to disseminate sensed data pushes it out to everyone. As there is no cross traffic, this approach is characterized with low contention. However, the drawback is that everyone might not be interested

Fig. 10.2 The push-based dissemination approach

Fig. 10.3 The pull-based dissemination approach

in the same data making an inefficient usage of network resources. The push-based approach is optimized for many receivers and few senders, Fig. 10.2.

The idea of *pull-based* dissemination is based on the *request-response* model and it is useful for user-specific data, e.g., webpage request. Motivated by application performance requirements, algorithms for dissemination are developed to match them to application requests like *one-phase pull* (an algorithm optimized for many senders and few receivers) and *two-phase pull*. Due to lots of cross traffic, this approach is characterized with contentions, collisions, and interference. The pull-based dissemination approach is given in Fig. 10.3.

The *flooding* approach is based on broadcasting the generated and received data to neighbors and usually everyone participates in the dissemination. This method is good for delay sensitive applications. However, the problem of broadcast storm arises. Techniques for avoiding it are based on simple forwarding, which can be timer based and hop limited, and Map-based/Geographic forwarding, which can be based on directed flooding or aggregation.

The *relaying* approach is based on the idea that the relay node forwards the data to the next hop and so on. The decision on which node to be selected as a next hop (relay neighbor) is either based on simple forwarding, by selecting the node which is farthest from the source, or is based on Map-based/Geographic forwarding which selects the node that is closest to the destination.

The architecture and the communication in VSNs are not necessarily only V2V or only V2I/I2V. When there is a hybrid architecture, i.e., V2V and V2I/I2V at the same time, the communication and the delivery of data between vehicles and among vehicles and RSUs must be improved. Additional data dissemination approaches can be obtained by combining the basic methods previously mentioned leading to *opportunistic*, *geographical*, *peer-to-peer,* and *cluster-based* dissemination. Their characteristics are briefly elaborated in the following paragraphs.

Opportunistic dissemination. The idea of opportunistic diffusion is that the messages are stored in each intermediate node and forwarded to every encountered node till the destination is reached. Therefore, the delivery ratio is improved. However, this kind of mechanism is not suitable for applications which are not delay tolerant.

Geographical dissemination. Due to the fact that end-to-end paths are not constantly present in a VSN, a geographic dissemination is used by sending the message to the closest node toward the destination till it reaches it.

Peer-to-peer dissemination. In a P2P solution, the source node stores the data in its storage device and does not send it in the network till another node asks for it. This solution is used and proposed for delay tolerant applications.

Cluster-based dissemination. For a better delivery ratio and to avoid or reduce broadcast storms, a message has to be relayed by a minimum of intermediate nodes to the destination. In order to do so, nodes are organized in clusters in which one node (a clusterhead) gathers data in its cluster and sends it after to the next cluster. The advantage of the cluster-based solutions is the lower propagation delay and high delivery ratio with bandwidth equity also. Avoiding the broadcast storm problem can be achieved by using a distributed clustering algorithm to create a virtual backbone that allows only some nodes to broadcast messages. An example of this approach is given in [27].

This section elaborated on some of the most relevant VSNs' basic mechanisms. These mechanisms provide the cornerstones of the VSNs' operation and management and clearly distinguish the VSNs as an intriguing networking paradigm. The following section will give a greater insight into the VSNs' protocol stack details and the layering architecture VSNs usually obey.

10.4 VSNs Layering Architecture

The layering architecture that most VSNs employ is shown in Fig. 10.4. The architecture differs according to the type of the deployed application, i.e., whether it is a *safety* or a *non-safety* one. Safety applications require a reliable transport protocol, suitable for real-time operation, and use the specifically tailored vehicular network layer, while the non-safety applications use the standard TCP/IP model. Finally, same physical and data link layers are used for both safety and non-safety applications. These two layers are based on different DSRC (Dedicated Short Range Communication) standards [13, 28].

The DSRC standards are the fundamental part of the V2V communication and, as such, a vital part of VSNs as well. In VANETs, the DSRC standards include the physical layer, the data link layer and the application layer where the first two serve as the foundation for VSNs. Similar standards have been standardized in Europe, USA, and Japan by different standardizing bodies as shown in Table 10.1.

Despite the number of different standards, IEEE 802.11p is slowly becoming the de facto DSRC standard worldwide being the most advanced and most suitable of all other

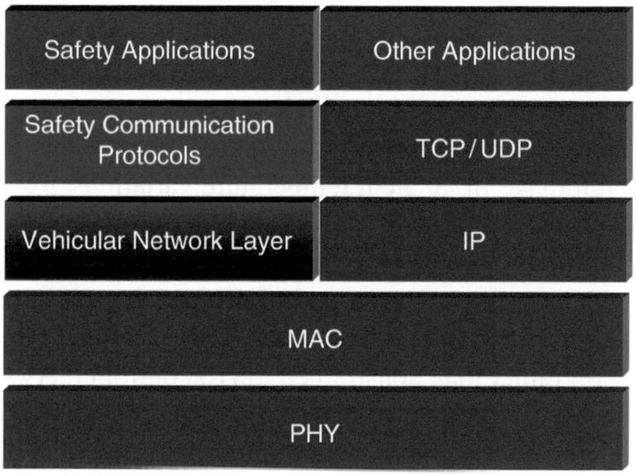

Fig. 10.4 Layering architecture in VSNs

Table 10.1 Comparison of various VSN-related standards worldwide

Country	Japan	USA	Europe
Name	ARIB-STD-T75	IEEE 802.11p	DSRC by CEN
Band	5.770–5.850 GHz	5.850–5.925 GHz	5.795–5.805 GHz
Duplex	OBU: half-duplex RSU: full-duplex	Half-duplex	Half-duplex
Communication system	Active	Active	Passive
Data rate	2 Mbps	3–27 Mbps	DL: 500 kbps UL: 250 kbps
Channels	7	7	4
Channel separation	5 MHz	10 MHz	5 MHz
Coverage	30 m	1,000 m	15–20 m
Modulation	2ASK (1 Mbps) 4PSK (4 Mbps)	OFDM	OBU: 2PSK RSU: 2ASK
Wireless access	TDMA–FDD	SDMA	TDMA–FDD

proposed standards. It is a WLAN standard very similar to IEEE 802.11a, but optimized for a vehicular environment. More on DSRC standards can be found in [13, 28].

Based on the previously elaborated basic mechanisms and the VSN layering architecture, the following section will introduce the VSN system implementation issues. It will pinpoint many practical aspects that use the elaborated mechanisms in order to provide VSNs' practical deployment.

10.5 VSN System Implementation

The popularity of VSNs emphasizes the practical aspects in their implementation. This section addresses possible implementation solutions in VSNs by identifying the necessary equipment for a VSN system structure deployment. In addition, it discusses various sensors and technologies and links them with examples of implementation.

10.5.1 Node Configuration

The architecture of a vehicle sensor system is determined by the specific environmental conditions and requirements it operates in. Determining factors comprise high range of ambient temperature variations, excessive dust and moisture content, high level of shock and vibrations, specific battery power supply, mobility, etc. Taking into consideration the rapid development of different vehicle specific sensors as well as the increasing demand for fast collection, processing, storage, and transmission of measured data, VSNs require the use of embedded computing platform rather than constrained micro controller. As shown in Fig. 10.5 [29], a vehicle sensor device contains a processor, a memory, a radio transceiver and antenna, a power source and an input/output interface that allows the integration of external sensors into the wireless device.

The central processing unit is a box computer, fan less and diskless design, small form factor, incorporating some of the Intel's "M" or "Atom" high speed and low power processors, powered by Windows Embedded or Windows CE platform. Providing a multitasking operating system, different node operations can be performed as separate threads or processes. For example, while one thread is employed in collecting and transforming the sensor data, the other can be dedicated to communication with other nodes in the network. Its robust design and low power requirements make this type of computers ideal for VSNs. Specially designed Vehicle Software Package (VSP) is in charge of reading raw sensed data, processing them, and generating summaries. Summaries include context metadata (location, timestamp etc.) and features extracted by local filters. For instance, VSP includes a filter that can determine position of the vehicle, vicinity to other objects, etc. Moreover, VSP commands the storage of both raw data and summaries in two local databases. Summary

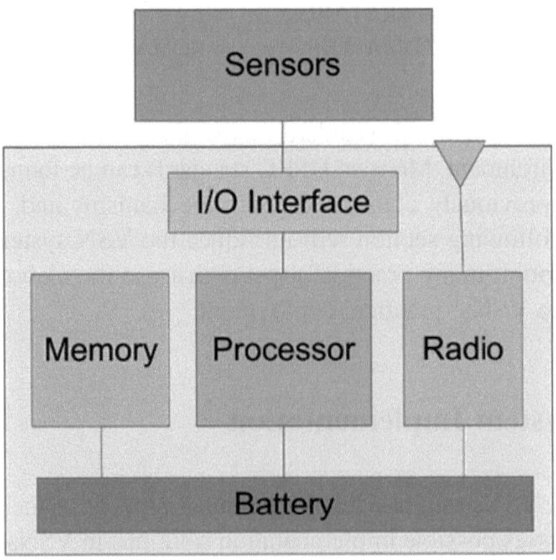

Fig. 10.5 VSN node (device architecture)

generation rate and size are crucial for device performance. Other characteristics of the suggested fan less box computer include Large Central processing memory (up to 4 Gb), connectivity (10/100BaseT LAN, RS232/485, USB), Peripherals (Keyboard/mouse, video VGA/LCD, parallel interface, IDE interface, audio/video interface), incorporated radio transceiver board, GPS possibility, fast multi channel A/D interface, FLASH type of internal and external memory, etc.

The radio transceivers contain all necessary circuitry to send and receive data over the wireless media: modulator and demodulator modules, digital to analog and analog to digital converters, low noise and power amplifiers, filters, mixers and antenna are among the most important components. Radio transceivers usually work half-duplex as simultaneous transmission and reception of data over the wireless media are impractical. Transceivers can operate in different modes (usually transmit, receive, idle, and sleep operational modes). The sleep mode is a very important energy-saving feature in WSNs. Some transceivers offer additional capabilities, including several sleep modes that turn different parts of the hardware on and off. Commercially available transceivers for WSNs have different characteristics and capabilities. Normally, they work on three different frequency ranges: 400 MHz, 800–900 MHz, and 2.4 GHz.

Vehicle sensor devices are usually powered by vehicle starter battery externally attached to the node. Therefore, power saving is important but is not crucial when sensors are deployed in the vehicles. If sensors are deployed as roadside sensors, network lifetime depends on how wisely batteries are used. Using the microcontroller and transceiver sleeping modes is crucial for long operation of the WSN working outdoor.

Finally, vehicle sensors devices are equipped with sensor boards, which contain application-specific sensors. The variety of sensors and sensor boards that can be directly interfaced with the wireless sensor device is very large. Temperature, air quality, pressure, magnetometers, light, acoustic, and accelerometers are just a small sample of the types of commercially available sensors. This interfacing flexibility is the cause of the wide popularity of WSNs as they serve as a general platform to solve practical problems in VSNs and in many other application domains.

Having discussed the general VSN node architecture, the following subsection will classify and elaborate on the actual sensor equipment being used in VSNs.

10.5.2 Sensor Equipment

Sensors have a key role in the VSN system structure. They should be accurate, reliable, and provide fast response. Depending on the location where they are deployed, VSN used sensors can be broadly divided into *roadside* sensors and *intra-vehicle* ones. Roadside sensors comprise different types, such as magnetic sensors, microwave radars, infrared, weather sensors, smog detectors, video detection, and others. They are used to provide information about the speed, direction, quantity of vehicles per unit time etc. As for the intra-vehicle sensors, there are distance sensors, speed sensors, etc., which measure traveled path, tilt sensing, operation time, load pressure, load temperature, load humidity, load weight and density, and many other parameters. This section will provide greater insight into the characteristics of the most relevant representatives of roadside and intra-vehicle sensors, respectively.

10.5.2.1 Magnetic Sensors

Wireless magnetic sensor nodes offer a very attractive, low-cost alternative to technologies such as inductive loops, video cameras, and radar for traffic measurement in freeways, urban street intersections, and presence detection in parking lots. They can provide information about speed and direction of traffic, quantity of vehicles per time on a stretch of pavement or just reliable presence or absence of a class of vehicles. Appealing to the fact that almost all road vehicles have significant amounts of ferrous metals in their chassis (e.g., iron, steel, nickel, cobalt, etc.), magnetic sensors are a good candidate for detecting vehicles. They usually use the disturbance of the magnetic field of the earth in order to determine the presence or absence of a vehicle.

Figure 10.6 shows a good graphical example of the lines of flux from the earth between the magnetic poles and the bending they receive as they penetrate a typical vehicle with ferrous metals. As the lines of magnetic flux group together (concentrate) or spread out (deconcentrate), a magnetic sensor placed nearby will be under the same magnetic influence the vehicle creates to the earth's field.

Most commonly used magnetic sensors are Anisotropic Magneto-Resistive sensors or AMRs [30]. AMRs are directional sensors and provide only an amplitude response to magnetic fields in their sensitive axis. By combining AMR sensors into two or three axis configurations, a two- or three-dimensional measurement of the magnetic fields passing through the sensor is possible with excellent linearity. For AMR sensors, the sensor-resistive elements are oriented as resistive "Wheatstone bridge" that varies resistance slightly as the magnetic field changes upon each element. There are several methods in which AMRs can be used. The following text describes them in detail.

Vehicle detection signature. As vehicles come near the sensor, there is a shift from the earth's magnetic field levels. Since the natural earth's magnetic field would bias the sensors with a slight negative voltage output, increasing flux concentration would

Fig. 10.6 Earth's magnetic field through a vehicle

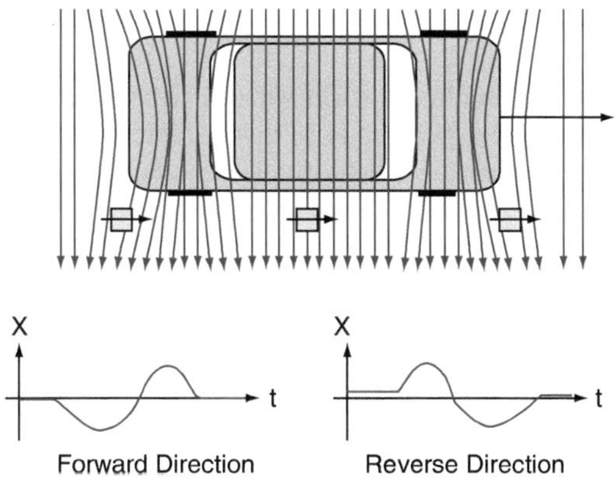

Fig. 10.7 Vehicle detection signature

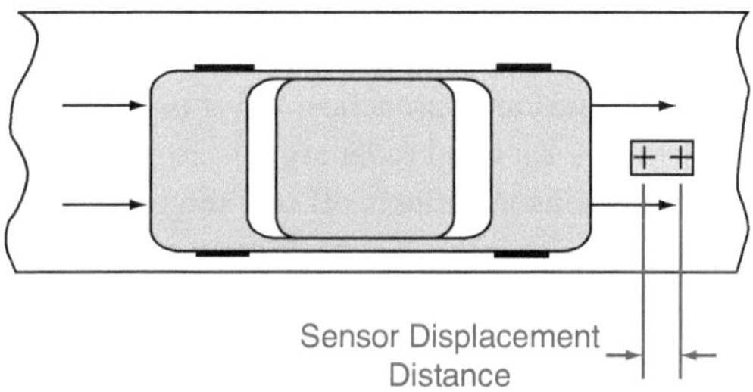

Fig. 10.8 Vehicle direction sensing

further lower the voltage and decreasing concentration would raise the voltage, Fig. 10.7 [28]. For most applications, the amplitude and direction of voltage shift are not important, but the detection of a significant shift in output voltage is of greater importance.

Vehicle direction sensing. This method involves two sensors displaced at a small distance apart, Fig. 10.8 [30]. The intention is that a vehicle in motion would create the same signature, but displaced in time. The nearer sensor will hit the detection threshold a fraction of a second before the front sensor in forward traffic. With a known displacement distance and a reasonably precise time measurement between thresholds detections, a speed computation can be made as well.

One of the biggest concerns for accurate vehicle detection system design is the reception of false presence of vehicles (i.e., "falsing"). Even though vandals may toss magnets onto the sensor areas and jeopardize the vehicle detection process, it is the nature made stimuli or adjacent lane vehicles that comprise most of the falsing concerns.

10.5.2.2 Intra-vehicle Sensors

Most common representatives of the intra-vehicle sensors are the *speed* and the *distance* sensors. This subsection elaborates on them in greater details pinpointing their most prominent characteristics.

Speed Sensors

A number of different technologies (fifth wheels, radar, microwave, optical, GPS) exist to measure the vehicle speed and use that parameter to calculate the distance traveled. Generally, speed sensors can be divided into two groups, i.e., *wheel-contact* and *non-contact sensors*. The second ones have much better performances, are easier to implement, and are much more recommended by today's security standards.

An interesting application of a vehicle's speed sensors is evaluation of the braking performance, i.e., measuring the total distance traveled by a vehicle during a stop. The method of measuring this distance needs to be accurate and repeatable in order to detect small variations in the braking performance. Measurements like this are suitable for testing performances of different types of speed sensors and also contribute to the non-contact sensors superiority. Figure 10.9 represents the process of non-contact speed measurement used by the Doppler radar speed sensor.

The Doppler radar is named after the Doppler principle, which explains the frequency shift associated with energy waves reflected by or emanated from a moving body. A familiar example of a Doppler shift is the change in pitch in the sound of a passing car – higher as the car approaches; lower as it leaves. The measurement process is done as follows. A Ka band radar signal (26.5–40 GHz) is transmitted at a specific frequency by the sensor, reflects off of a target (or targets), and returns to the sensor (see Fig. 10.9). If either the sensor or the target is moving relative to one another, the signal will be shifted in frequency when it returns to the sensor. The fundamental Doppler frequency shift is given by:

$$F_{\mathrm{d}} = 2 \cdot V \cdot \frac{F_0}{c} \cdot \cos\theta \tag{10.1}$$

where F_{d} is the Doppler frequency shift (Hz), V is the velocity, θ is the offset angle of the sensor relative to the direction of the target motion, c is the speed of light and $F_0 = 35.5 \pm 0.1$ GHz (Ka band).

Equation (10.1) is used to accurately obtain the velocity of the vehicle by measuring only the Doppler frequency shift since all other variables are known in advance.

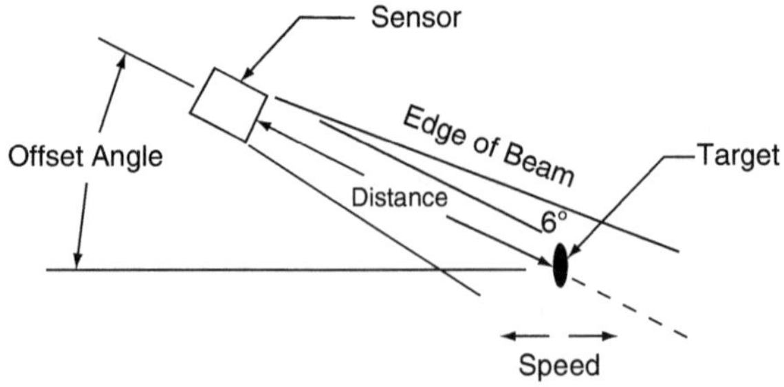

Fig. 10.9 Non-contact speed measurement

Distance Sensors

Distance sensors measure the distance from the sensor to a target object. Based on the measurement method, their classification is given in Table 10.2. Special interest for intra-vehicle implementation lies in the radar, laser, and ultrasonic sensors. The distance is determined by sending out ultrasonic pulses and listening for the returning echoes reflected from the target. Thus, the use of the Doppler effect is prevalent again as in the case of speed measurement sensors. The time from transmissions to return of an echo is the basis of obtaining the distance measurement. Nowadays, instead of sound (ultrasound), this method is used with electromagnetic waves and light or laser. It is important to understand that the wave frequency takes an important role in the distance measurement.

Table 10.3 gives a comparison of three ultrasonic sensors. As anticipated, an increase in device dimensions leads to an increase in their performances like range,

Table 10.2 Distance sensor classification

Distance sensors	
Long distances	**Short distances**
• Radar	• LVDT
• Ultrasonic	• Triangulation laser
• Long distance laser	• Profile scanner laser
• By cable	• Potentiometer
• Magnetoresistive	• Magnetorestrictive
	• Digital

Table 10.3 Distance sensor's characteristic

	SR50	SA6A	SRF235
	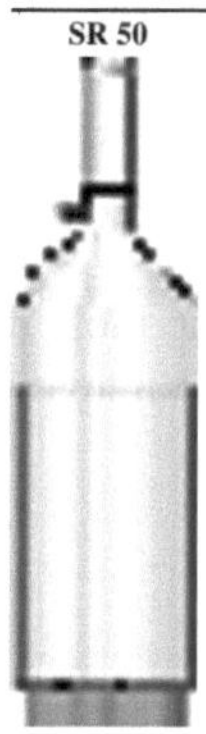		
Power consumption (mA)	2–250	10	25
Measurement range (cm)	50–1,000	20–200	10–120
Response time (ms)	60	48–90	?
Dimensions (mm)	310 × 75 × 75	84.5 × 50 × 29	34 × 20 × 19
Weight (g)	1,300	270	6

response time, etc. Sensor's dimensions are a significant factor in the process of system building; therefore, a trade-off in size-performances is often needed.

This section focused on the system implementation aspects of VSNs. It discussed the relevant node architecture and the characteristics of the most popular sensors used in VSNs today. Further development of VSNs and their practical implementations often require integration of different technologies that are able to sense, process, and send data within a dynamic networking context. The following section will focus on a technology that is not novel per se, but its usage within the VSN concept allows an evolution toward an ubiquitous and pervasive VSN environment capable of autonomous and user-transparent operation.

10.6 Evolution Toward RFID

The increasing interest from both academia and industry for VSNs yields evolution of the VSN concept toward novel technologies that closely resemble the peculiarities of the vehicular environment. These new technologies must incorporate elements from mobile computing and automatic data collection systems and enable self-configuration, lightweight implementation, and integration with other sensor networking technologies. They should strengthen the sensing capabilities of the vehicular environment allowing for more efficient VSNs' operation. One of the most prominent candidate technologies for integration within the VSN paradigm today is the RFID.

RFID stands for Radio Frequency Identification. It is a technology that incorporates the use of electromagnetic or electrostatic coupling in the radio frequency portion of the electromagnetic spectrum to uniquely identify an object, animal, or a person. This technology is similar to the bar code identification systems present in retail stores every day, but the advantage of the RFID technology is that it does not require direct contact or line-of-sight scanning [31]. The system's advantages are: multiple sensor readings, faster data acquisition, interoperability, reusability, and flexibility, which offer great potential and can be very useful in a VSN implementation. Therefore, this section will elaborate the most important characteristics of the RFID technology paving the evolution of the integration between VSNs and RFID.

10.6.1 RFID System

A basic RFID system consists of three components, i.e., an antenna and a transceiver (often combined into one *RFID reader*), and a transponder (*RFID tag*), which is electronically programmed with unique information and a *database* [32]. The electronics in the reader uses an external power source to generate the signal that drives the reader's antenna. In turn, the antenna creates the appropriate radio wave. This radio wave may be received by an RFID tag, which, as a result, "reflects" some of the energy it receives in a particular way. While this reflection occurs, the RFID reader is also acting as a radio receiver so that it can detect and decode the reflected signal in order to identify the tag. The reader is larger, more expensive, and it needs more power compared to the RFID tag. There are a number of different types of RFID systems, but one basic classification is based on the power source used by the tag [33].

10.6.2 RFID Tags

An RFID tag is a microchip combined with an antenna in a compact package. The packaging is structured to allow the RFID tag to be attached to an object that is to be tracked. The tag's antenna picks up signals from the RFID reader and then returns the signal, usually with some additional data (like a unique serial number or other customized information). Some RFID tags can be very small, the size of a large rice grain, while others may be the size of a small paperback book. In RFID systems, the tags that hold the data are divided into two different types, i.e., *active* and *passive* tags.

10.6.2.1 Passive Tags

Passive tags have no battery and do not initiate communication. Instead, they draw power from a reader, which sends electromagnetic waves that induce a current in the tag's antenna and powers the microchip's circuits. Consistent with any applicable authorization, authentication, and encryption, the tag will then respond to the reader, sending via radio waves the data stored on it.

The advantage of a passive tag is the fact that these tags have a useful life of 20 years or more and are typically much less expensive to manufacture. Also, the tag is of very small size (rice grain proportions). These tags have almost unlimited applications in consumer goods and other areas. But, there are some disadvantages of the passive tags. For example, the tag can be read only at very short distances, approximately 30 ft. This greatly limits the device for certain applications. Also, it may not be possible to include sensors that can use electricity for power. The tag remains readable for a very long time, even after the product to which the tag is attached has been sold and is no longer being tracked [34].

Semi-passive tags. These tags do not initiate communication with readers (like the passive ones); however, they do possess batteries providing on-board power used for storing information, e.g., ambient temperature. The semi-passive tags can be combined with other sensors to create "smart dust", tiny wireless sensors that can monitor environmental factors. For example, a vineyard might use smart dust to measure incremental weather changes that could critically affect grapes [32].

10.6.2.2 Active Tags

An RFID tag is an active tag when it is equipped with a battery that can be used as a partial or a complete source of power for the tag's circuitry and antenna. Some active tags contain replaceable batteries for years of use, others are sealed units. It is also possible to connect the tag to an external power source. The major advantages of active RFID tags are that it can be read at distances of 100 ft or more, greatly improving the utility of the device. However, active RFID tags are typically more expensive and physically larger, which may limit applications [34].

Both tagging technologies, i.e., active and passive, are needed for RFID to release its full potential. The cost savings afforded by passive tags makes RFID tagging possible at a much lower price point than would be possible with active tags alone. On the other hand, active tags add functionality not possible with passive tags.

RFID tags can be also classified according to the memory capabilities. Namely, *read-only* tags have a unique ID code which is permanently stored on the tag, while the *read/write* tags are allowing a user to change the ID and add additional data to the tag's memory. In addition, these tags can be combined with a permanent tag ID and some storage space for the user's data. Passive tags typically have anywhere from 64 bits to 1 kB of non-volatile memory. Active tags tend to have larger memories with a range of, typically, between 16 bytes and 128 kB [32].

10.6.3 RFID Reader

An RFID reader or a scanning device typically contains a *module* (transmitter and receiver), a *control unit,* and a *coupling element* (antenna). They vary in size, weight, and power, and may be mobile or stationary. The reader has three main functions: energizing, demodulating, and decoding. In addition, readers can be fitted with an additional interface that converts the radio waves returned from the RFID tag into a form that can then be passed on to another system, like a computer or any programmable logic controller. Anti-collision algorithms permit the simultaneous reading of large numbers of tagged objects, while ensuring that each tag is read only once. This is any device, such as a laptop computer, an electronic scale head, or a handheld computer, that is capable of communicating with a transceiver and accepting the information from it. Usually, the RFID-tagged information is then stored in an RFID database used to tie electronic identity to production and management information, message the data, and share the information with others [35].

There are two main classes of RFID readers, i.e., *read-only* and *read/write*. The main difference is that the latter ones can write new information back to a tag that has been equipped with a read/write memory. The readers are becoming increasingly sophisticated, acting as gateways into the network-centric communication systems of modern enterprises by supporting communication protocols such as TCP/IP and network technologies such as DHCP, UDP/IP and Ethernet, or 802.11× (for wirelessly sending data back to the enterprise). Many models of readers are handheld devices and resemble the pricing guns or barcode scanners used in supermarkets. Readers can also be fixed in place (e.g., in doorways or at vehicle toll gateways) and even hidden, e.g., embedded into ceilings or walls [32].

10.6.4 RFID Frequency Allocation

RFID is fundamentally based on wireless communication, making use of radio waves, which form part of the electromagnetic spectrum (i.e., frequencies from 300 kHz to 3 GHz). It operates in an unlicensed spectrum space referred as ISM (Industrial, Scientific, and Medical). The exact frequencies that constitute ISM may vary depending on the regulations in different countries, Table 10.4.

Low-frequency (LF) tags operate at less than 135 kilohertz (kHz), which typically means short-range uses, like animal identification and anti-theft systems (RFID-embedded automobile keys). System that operates at 13.56 megahertz (MHz) is characterized as high frequency ("HF"), which means longer read range. Scanners can read multiple HF tags at once and at a faster rate than LF tags. The third frequency, Ultra-High Frequency ("UHF"), operates at around 900 MHz, and it can be read at longer distances, which outside the laboratory environment range between 3 and possibly 15 ft. However, UHF tags are more sensitive to environmental factors like water, which absorb the tag's energy and thus block its ability to communicate with a reader.

RFID systems based on LF and HF frequencies make use of near-field communication and the physical property of inductive coupling from a magnetic field. The reader creates a magnetic field between the reader and the tag and this induces an electric current in the tag's antenna, which is used to power the integrated circuit and obtain the ID.

Table 10.4 RFID frequency allocation

Types	Used frequencies	Comment
Low Frequency (LF)	125–134 kHz (Europe, North America and Japan)	• Short read range (<0.5m) • Less sensitive to interference • Less expensive
High Frequency (HF)	13.56 MHz (Europe, North America, Australia, and Japan at very similar power levels)	• Read range (<1 m) • Higher speed than LF
Ultra High Frequency (UHF) – non microwave	860–930 MHz (EU and USA)	• Read range (up to 3 m)
	950–956 MHz (Japan)	• Limited ability to read tags with high water or metal content
Ultra High Frequency (UHF) – microwave	2.45 GHz (used by IEEE 802.11b and 802.11g WLAN and Bluetooth systems)	• Offer the highest data read rates • Most expensive

The ID is communicated back to the reader by varying the load on the antenna's coil, which changes the current drawn on the reader's communication coil. RFID systems based on UHF and higher frequencies use far-field communication and the physical property of backscattering or "reflected" power. Far-field communication is based on electric radio waves. Actually, the reader sends a continuous base signal frequency that is reflected back by the tag's antenna. During the process, the tag encodes and modulates the signal to be reflected with the information from the tag (the ID).

The evolution of VSNs requires various systems, from embedded to fully Internet capable ones, to interoperate, adapt to various environmental conditions, and deliver a user-transparent and seamless vehicular environment communication. It inevitably leads to adoption of novel technologies and communication concepts that facilitate automatic information collection and dissemination under dynamic conditions. The RFID technology has the potentials to deliver these necessary prerequisites for future generation VSNs. Its characteristics allow it to be easily integrated into a vehicular environment. Nowadays examples comprise fleet management systems [36], the RUNES project defined integrated heterogeneous embedded systems [37] etc. However, the development of the RFID technology must be viewed with caution and jointly with VSNs to allow the evolution. Otherwise, it may devolve into proprietary-based single-use applications that limit its potentials.

10.7 VSNs Applications

After scrutinizing the most important aspects of VSNs, this section will focus on classification and analysis of the most prominent VSNs applications in use today. The applications are the driving force of vibrant research in all technology fields and the VSNs are no exception to this rule.

VSNs' applications can be broadly divided into:

- *Safety applications* (potential killer applications)
- *Non-safety applications*

Safety applications are used for traffic accident prevention and life's salvation, whereas the non-safety applications are a step forward toward ambient intelligence in a vehicular environment. The data used for VSN applications is gathered mostly by sensors capable of detecting distance, position, and speed. This subsection will pinpoint some of the most prominent and interesting VSN applications today conveying the previously mentioned classification.

10.7.1 Safety Applications

The VSNs safety applications' target is the "safety margin", i.e., the time in which a potential accident is detected before it may actually occur. The objective is therefore to inform the drivers about a potential risk sufficiently in advance to avoid emergency maneuvers and, thus, extend the safety margin. Relevant examples of this type of applications are further discussed.

Pre-crash sensing/warning. Pre-crash sensing system application uses a millimeter-wave radar that can determine object's position, distance, and speed in order to minimize injury and damage in the event of a collision. The sensed information goes into a pre-crash safety computer that takes into account relative velocities, distance, and other factors to determine whether or not a collision can be avoided. If not, it notifies the seatbelt control computer, which tightens up any slack in the seatbelts to maximize the initial restraint on the driver and passengers at the moment of the crash [38].

Traffic Display Monitoring (TDM) [39]. TDM's Digital Display Radar Trailers are designed specifically to promote driving safety on virtually any public or private roadway. Use of TDM's Digital Display Radar Trailers produces numerous notable benefits such as fewer fatalities, fewer injuries, less property damage, fewer vehicle-related lawsuits, possibly lower insurance premiums, etc. They are easily towed to sites experiencing speed-related problems or where accident reduction is a major priority. Setup is uncomplicated and a unit can be operational in a matter of minutes. Speed is shown on a highly visible, amber-colored LED display (Fig. 10.10).

Safe lane change maneuvers. This application allows vehicles in the blind spots and vehicles that are intending to change lanes to be detected in advance. The information is then disseminated to all drivers of relevant vehicles.

Road departure prevention. The road departure prevention application allows for vehicles to receive information on recommended speed from road side infrastructure. The recommended speed is calculated according to road geometry, surface status, and traffic conditions. In addition, vehicles equipped with sensors measuring road friction can also communicate to other vehicles the presence of slippery roads.

Fig. 10.10 TDM's top-of-the line [39]

Cooperative maneuvering (e.g., highway merging). In this application, vehicles calculate in real time their relative position and trajectories. Then, when a risky situation and/or a potential collision is detected, drivers of relevant vehicles can be promptly warned.

Cooperative tunnel safety. This application is specifically designed for road tunnels. It relies on the roadside infrastructure that informs vehicles about recommended speed and safety distance. As a result, the safety margin is calculated on the basis of the state and typology of the vehicle.

Hazard and incident warning. This application targets transmission of warning messages to vehicles arriving in an area where an accident has just occurred. The message can be issued from the roadside infrastructure or from other vehicles and may include type of hazard, current location, and previous positions, speed, direction, etc.

Safe urban/extra urban intersections. This application requires a very precise computation of the vehicles trajectories and local digital maps of the intersections. The infrastructure delivers information to the vehicles to recognize dangerous situations in time [4].

Railway safety [40]. Overheated axles and locked brakes are the main immediate causes of potential failures in railway operation. As a result of this, the axle casting may break causing the train to derail. Furthermore, locked brakes represent potential problems, e.g., overheating releases wheel rims or may cause wheel disks to break. Locked brakes may also result in fires leading to the formation of treadles spots. The railway safety system application [40] can identify all bearing types and braking systems used in international railways at speeds of up to 500 km/h and reliably detect potential areas for failure. In addition to being competitive, the system is modular, allowing for rapid and low-cost maintenance. The core component of the system is an infrared optical sensor that conducts redundant row-scans of the bogie at widths of up to 120 mm. The system has an integrated signal processor for fast and accurate processing of measured values. The infrared sensors are designed to be plug-in modules, integrated into a hollow sleeper, and include all the microelectronic components needed to process the measured values. They are precisely positioned by pre-adjusted module frames and can be fully replaced for servicing within just a few minutes without distorting the measurement geometry. Servicing does not involve any adjustment whatsoever at all! The measured values are processed and sent via a secure high-speed data transmission line to the Service & Communication Terminal (SCT). Integrated digital electronic components perform the task of processing the recorded thermal images as well as all the controlling functions for the sensor module. A detailed temperature profile with 450 scanned points per axle with a diameter of 25 cm can be recorded at train speeds of 500 km/h (this corresponds to 1,125 scanned points at 200 km/h). Built-in environmental sensors are used to take into account all the relevant vectors (e.g., moisture and temperature) during this process. Ongoing auto-diagnostics of all system components as well as automatic service request indicators have been included for high reliability and economic operation.

Electronic train management system [41]. This system integrates new technology with existing train control and operating systems providing a safety overlay, working in conjunction with existing train operations, patterns, and rules. The system is developed to prevent track authority and speed limit violations and ultimately reduce the potential for train accidents. It uses wireless communication technology to transmit train movement–related information between locomotives and Network Operation Center (NOC) computer systems. Information about the train comprises the track database, speed limits, movement authorities, and work zones loads into the on-board computer. The system also integrates and interlocks the train information with data from a GPS to create, in effect, a safety overlay for train operations. It considers movement-related information, like the movement authority limits, speed limits and work zones through both WiFi and radio frequency communication networks and displays it in a graphic interface inside the locomotive cab. At the same time, the system pings wayside devices along the track through a low band wireless communication network checking for broken rails, proper switch alignment, and signal aspect information. As the train moves, the computer constantly calculates a warning and braking curve based on all mentioned information. The warning curve provides a margin of safety to warn the crew so that they can slow or stop the train before it reaches the braking curve boundary. For example, as the train's warning curve reaches the limit of the train authority, the train crew receives a warning signal on the on-board computer to stop. If the locomotive engineer does not stop the train before the stopping distance is reached, the system will automatically use enforcement braking and stop the train with a full service brake application.

Railway bridge monitoring [42]. Railway systems consist of huge number of bridges being used over several decades. It is important to have a system to monitor the bridge health and report when and where maintenance operations are need. The railway bridge monitoring systems usually implement battery-operated wireless sensor nodes. The use of wireless transceivers and battery eliminates the need of having to lay cable to route data or power to the various sensors that are spread on the bridge. The choice of a battery-operated wireless nodes is key for low-energy consumption so that the maintenance requirements are kept to a minimum. The nodes need to sleep most of the time to conserve power, but they also need to be ready for taking accelerometer measurements when there is a passing train. This mechanism of work is called sleep-wakeup mechanism. On detecting a train, nodes collect vibration data. The collected data then has to be transferred to a central location, where it will be used for analyses. The passing trains can be used for the data transfer.

10.7.2 Non-Safety Applications

The non-safety VSN applications do not target safety per se, but allow for construction of intelligent transportation and traffic management systems. There are numerous examples in this group ranging from different traffic parameter monitoring (enabling traffic efficiency), green light optimal speed advisory, automatic door/gate opening, parking meters, railroad crossing control to unmanned vehicle operation. Some of the most relevant representatives of the non-safety VSN applications are discussed below.

Drive-through retail (e.g., banking, fast-food). The drive-through retail application for vehicle detection and tracking has been exploding in recent years, and the amount of data precision is forcing technology usage. It is typically used to track customer vehicles from order taking, to payment, to service delivery positions.

Parking lot space detection and parking meters [30]. Parking space and parking meter applications have similar constraints for low cost and reliable presence/absence detection of stationary vehicles. Some more modern parking meter designs also contain special sensors. These sensors use laser lights to determine when the parking spot has been vacated. When the parking meter determines the spot is empty, the time remaining on the clock automatically returns to zero. This prevents other people from using the time remaining on a parking meter after the spot has been vacated. The special sensor also ensures that cars are removed from areas that have a limit on the amount of time a person can remain parked in one spot. Figure 10.11 shows a typical parking meter arrangement with three sensor location.

Unmanned railway operation. The first unmanned train operation, using the same infrastructure in the same time with the ordinary trains, was set out in the underground stations in Nurnberg [43]. To allow for automatic train control, a network of several computers is installed in a tunnel under Nurnberg. Using light waves and wireless, the system is in constant connection with computers and trains without the need for a driver. Employers in control centers monitor all processes and are always prepared to intervene if there is a necessity. Platforms are monitored with cameras and special systems for tracking. For example, if someone or something falls on railway tracks, the system automatically will stop all trains in that particular sector of the network. Also, if an infrared sensor registers that something makes a pressure on the door, then the train will not leave the platform. In the meantime, the passengers are not aware that they are in train without a driver and can enjoy the sight in front of them, not blocked with the driver cabin.

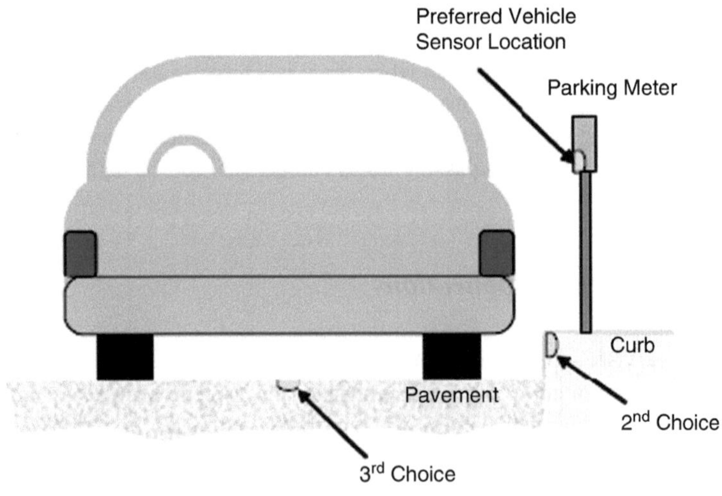

Fig. 10.11 Parking meter sensing

This section covered some of the most relevant VSNs applications today. As the field is very vibrant, numerous novel applications are being constantly introduced. They all drive the research and the implementations toward an ambient vehicular network design able to provide safety and infotainment in a user-transparent and seamless way.

10.8 Conclusions

VSNs are increasingly attracting the interest of both academia and industry as manufacturers constantly equip vehicles with a plethora of sensing possibilities. The ability to use the installed vehicle sensor equipment to sense the vehicular environment and send this data to relevant entities yields research and implementations that increase traffic safety or manage traffic situations in a cooperative (i.e., more intelligent) way. It contributes to the development of an embedded intelligence on the roads able to judge autonomously in the most efficient manner under various circumstances and for various applications.

The specifics of the vehicular environment (e.g., dynamic topologies, speed limitations, drivers' habits etc.) make its peculiarities very challenging. Large number of projects has been conducted in this area in the past few years, with the interest ranging from developing innovative telematics applications to defining an open architecture and processing multisensorial signals [14]. However, many issues are yet to be tackled. The development of the VSN technology is expected to be concentrated on improving the channel modeling, scalability, modulation and coding techniques, mobility and network management, routing mechanisms, etc. The possible VSN implementation studies clearly point that sensors must be small sized, independent of temperature and electromagnetic fields, interoperable, and capable of multiple sensor reading. Once all these issues are resolved, VSNs may become an integral part of future traffic management solutions. In addition, RFID can become fully integrated into current and emerging standards based automatic information collection systems. But, RFID must not be taken for granted as it can devolve into proprietary-based single-use applications that ultimately limit its potential.

This chapter gave an overview on the general VSN aspects and some of the most prominent VSN implementation issues. It should serve as an introductory text in the area giving an intriguing snapshot of the potentials of the VSN technology. Even though one can conclude with utmost certainty that VSNs will be one of the key components in future wireless and mobile networking solutions, the pace of their adoption as well as usage strongly depends on the actual research done in the area.

Acknowledgment The authors would like to thank Bisera Jankuloska, Milan Zahariev, Mihajlo Pavloski, Sanja Bonevska, Simona Temelkova, Natasa Paunkoska, Andrijana Jankova, Kiril Popadinski, Vanco Popov, Marijana Neskovska and Makedonka Gudzirova, undergraduate students in telecommunications (2008/2008) at the Faculty of Electrical Engineering and Information Technologies (FEEIT) – Skopje, for their valuable contributions and dedicated work on the usage scenarios within the ProSense project [44].

References

1. Nekovee M (2005) Sensor networks on the road: the promises and challenges of vehicular adhoc networks and vehicular grids. In: Proceedings of the workshop on ubiquitous computing and e-Research, Edinburgh, UK, May
2. Lim KW, Jung WS, Ko Y-B (2008) Multi-hop data dissemination with replicas in vehicular sensor networks. IEEE VTC Spring 2008, Singapore, May 2008
3. CVIS, EU FP6 Integrated Project (2006–2010). Available at http://www.cvisproject.org
4. SAFESPOT, EU FP6 Integrated Project (2006–2010). Available at http://www.safespot-eu.org
5. Mobile Millenium project (2008–2009). Available at http://traffic.berkeley.edu/index.html
6. GeoNet, *EU FP7 STREP* (2008–2010). Available at http://www.geonet-project.eu
7. Torok A, Laborczi P, Gerhath G (2008) Spatially constrained dissemination of traffic information in vehicular ad hoc networks. IEEE VTC Fall 2008, Calgary, Canada, Sept 2008
8. Car-2-Car Consortium. Available at http://www.car-to-car.org (last accessed date 7.09.2010)
9. eSafety Support Forum. Available at http://esafetysupport.org (last accessed date 7.09.2010)
10. Bakker D, Gilster DM, Gilster R (2002) Bluetooth end to end. Wiley, New York
11. Tsai H-M, Saraydar C, Talty T, Ames M, Macdonald A, Tonguz OK (2007) ZigBee-based intra-car wireless sensor network. IEEE ICC2007, Glasgow, Scotland, June 2007
12. Ghavami M, Michael L, Kohno R (2007) Ultra wideband signals and systems in communication engineering. Wiley
13. Bilstrup K (2007) Vehicular communication standards. Telematics valley conference, Göteborg, Sweden, Nov 2007
14. Strandén L, Uhlemann E, Strom EG (2008) State of the art survey of wireless vehicular communication projects. 15th world congress on Intelligent Transportation Systems (ITS), New York, Nov 2008
15. Akyildiz I, Wang X (2009) Wireless Mesh Networks, John Wiley and Sons
16. Djenouri D, Soualhi W, Nekka E (2008) Mobility models in vehicular ad hoc networks: the overtaking impact. Proc Mediterr J Comput Netw 4(2):62–70
17. Bai F, Sadagopan N, Helmy A (2003) IMPORTANT: a framework to systematically analyze the Impact of mobility on performance of routing protocols for ad hoc networks. IEEE INFOCOM 2003, San Francisco, CA, 30 March–3 Apr 2003
18. Johnson DB, Saha AK (2006) Modeling mobility for vehicular ad hoc networks. In: Proc. of the 3rd international ACM workshop on Vehicular Ad-hoc NETworks VANET'06. ACM Press, New York, pp 30–39
19. Mahajan A, Potris N, Gopalan K, Wang AIA (2007) Urban mobility models for vanets. IEEE WCNC 2007, Hong Kong, Mar 2007
20. Choffnes DR, Bustamante FE (2005) An integrated mobility and traffic model for vehicular wireless networks. In: Proc. of the 2nd international ACM workshop on Vehicular Ad-hoc NETworks VANET'05. AMC Press, New York
21. Karnadi FK, Mo ZH, Lan K-c (2007) Rapid generation of realistic mobility models for VANET. IEEE WCNC 2007, Hong Kong, Mar 2007
22. Krajzewicz D, Behrisch M SUMO – Simulation of Urban Mobility. User documentation 7 Sep 2010. Available at http://sumo.sourceforge.net/ (last accessed date 7.09.2010)
23. Gorgorin C, Gradinescu V, Diaconescu R, Cristea V, Iftode L (2006) An integrated vehicular and network simulator for vehicular ad hoc networks. In: Proceedings of 20th European Simulation and Modeling Conference, Oct 2006
24. Jaap S, Bechler M, Wolf L (2005) Evaluation of routing protocols for vehicular ad hoc networks in city traffic scenarios. In: Proc. of the 5th international conference on Intelligent Transportation Systems (ITS) Telecommunications, Brest, France, June 2005
25. Stojmenovic I, Giordano S, Blazevic Lj (2002) Position based routing algorithms for ad hoc networks: a taxonomy. White Paper. Available at http://www.site.uottawa.ca/~ivan/routing-survey.pdf

26. Stojmenovic I (2002) Position based routing in ad hoc networks. IEEE Commun Mag 40(7):128–134
27. Salhi I, Cherif M, Senouci SM (2008) Data collection in vehicular networks. Autonomous and Spontaneous Networks Symposium, Paris, France, Nov 2008
28. Kudoh Y (2004) DSRC standards for multiple applications. 11th world congress on ITS, Nagoya, Japan
29. Labrador MA, Wightman PM (2009) Topology Control in Wireless Sensor Networks, Springer.
30. Application Note-AN218 (2005) Vehicle detection using AMR sensors. Honeywell International Inc. Available at http://www.honeywell.com
31. Landt J (2001) Shrouds of time: the history of RFID. AIM, Pittsburgh, PA
32. Ward M, van Kranenburg R (2006) "RFID: frequency, standards, adoption and innovation," JISC Technology and Standards Watch, May 2006. http://www.rfidconsultation.eu/docs/ficheiros/TSW0602.pdf (last accessed date 7.09.2010)
33. Hodges S, Harrison M (Oct 2003) Demystifying RFID: principles and practicalities. White Paper, Cambridge, MA
34. Atlas RFID Solutions Activer RFID vs. Passive RFID. Available at http://www.atlasrfidsolutions.com/active-vs-passive.asp (last accessed date 7.09.2010)
35. RFID: Applications and Implications for Consumers. Workshop Report from the Staff of the Federal Trade Commission, March 2005. Information available at: http://www.ftc.gov/os/2005/03/050308rfidrpt.pdf (last accessed date 7.09.2010)
36. Advantech Co. (2006) Vehicle and fleet management solutions. Available at http://www.advantech.com
37. RUNES EU FP6 Integrated Project (2004–2007). Avaialble at http://www.ist-runes.org
38. Denso Corporation Technology. Pre-crash Safety System. Available at http://www.globaldenso.com/en/technology/product/electronics/files/pdf12_e.pdf (last accessed date 7.09.2010)
39. TDM 1-radar-speed-trailer-display. Available at http://1-radar-speed-trailer-display.com/home.htm (last accessed date 7.09.2010)
40. VAE Hot box and hot wheel detection system PHOENIX MB. Available at http://www.voestalpine.com/vae/en/products/railway_infrastructure/switchsystems/electronic_systems/hazard_notification/hoa_phoenix.html (last accessed date 7.09.2010)
41. Wabtec Electronic Train Management System. Available at http://www.wabtec.com/railroad/etms.asp (last accessed date 7.09.2010)
42. Chebrolu K, Raman B, Mishra N, Valiveti PK, Kumar R (2008) BriMon: a sensor network system for railway bridge monitoring. In: Proceedings of the 6th international conference on mobile systems, applications and services, Breckenridge, CO
43. Nuremberg Automatic U-Bahn, Germany (2008). Available at http://www.railway-technology.com/projects/neuremburgautobahn/
44. ProSense, *EU FP7 CSA project* (2008–2010). Available at http://www.prosense-project.eu

Part III
Case Studies

Chapter 11
FEEIT WSN Testbed: Effective System for Providing Emergency Situations' Prevention and Surveillance

Aleksandra Mateska, Vladimir Atanasovski, and Liljana Gavrilovska

This chapter presents possible applications of WSNs, along with the emerging technology of RFID, by elaborating on a deployed sensor network research testbed at FEEIT's (Faculty of Electrical Engineering and Information Technologies – Skopje, Macedonia) faculty premises. The installed system serves for development of demos/proof-of-concept in laboratory surroundings mainly in the area of emergency and surveillance. The testbed and the experience from the demo applications development are envisioned as an initial phase toward building an advanced large-scale setup and, further on, as a basis for comprehensive research. This chapter focuses on the process of testbed deployment, elaborates selected usage scenarios with appropriate hardware and software resources and their integration into a joint testbed platform.

The book chapter is organized as follows. Section 11.1 introduces the testbed and the selected usage scenarios. Section 11.2 describes the architecture and the implementation of the application for scenarios' monitoring. The process of testbed realization through different demo applications is described in Section 11.3. The lessons learned from the testbed deployment are briefly described in Section 11.4. Finally, Section 11.5 concludes the chapter.

11.1 FEEIT's Testbed

The FEEIT's testbed exploits Sun SPOT sensor technology [1], diverse sensors (speed, distance, motion, humidity, wind speed, sound, temperature, smoke, and light [2–10]), and RFID development kits [11–13]. Stemming from the common testbed goal, i.e., developing demo applications for emergency and surveillance, five usage scenarios are targeted:

A. Mateska (✉), V. Atanasovski, and L. Gavrilovska
Faculty of Electrical Engineering and Information Technologies, Skopje
e-mail: amateska@feit.ukim.edu.mk; vladimir@feit.ukim.edu.mk

L. Gavrilovska et al. (eds.), *Application and Multidisciplinary Aspects of Wireless Sensor Networks*, Computer Communications and Networks,
DOI 10.1007/978-1-84996-510-1_11,

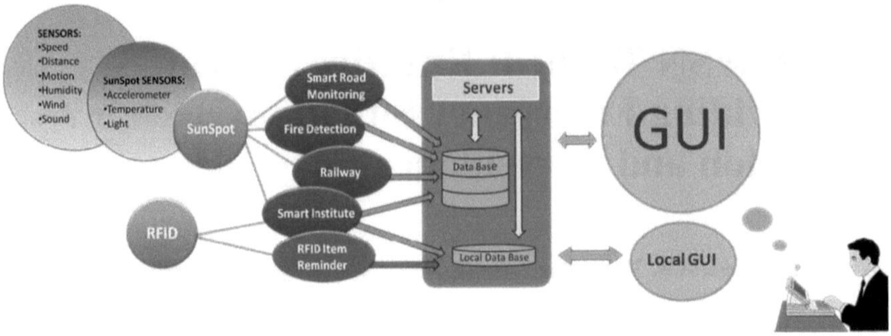

Fig. 11.1 FEEIT's testbed architecture overview

1. *Smart Institute*
2. *RFID Item Reminder*
3. *Smart Road Monitoring*
4. *Railway Sensor Networks: Secure Transport*
5. *Early Fire Detection*

Two of the scenarios, *Smart Institute* and *RFID Item Reminder*, refer to the usage of WSN and RFID for surveillance. Another two, *Smart Road Monitoring* and *Railway Sensor Networks: Secure Transport*, concern the safety issues in the traffic communications (vehicular and railway). The last one, Early Fire Detection, targets early detection of forest fire.

Applications for each of the selected usage scenarios are mostly performed in Java (*RFID Item Reminder* uses C-Sharp) and stored on a server that handles reprogramming and data logging, while providing a web interface for online scenarios monitoring and appropriate Graphical User Interface (GUI). The goal of the monitoring system is real time, interactive and remote sensors' monitoring, set up in different environments applicable in relevant use cases. The monitoring includes reading parameters from the sensors and their visualization. The general architecture of the FEEIT's testbed is depicted in Fig. 11.1.

The testbed further employs notification devices (GSM/GPRS gateway [14] and alarm) for events reporting to defined authorities. The following section presents the joint GUI and the developed application for the monitoring of the sensors' activity.

11.2 Application for Monitoring the Sensors' Activity

The application for monitoring sensors' activity resembles the testbed architecture overview depicted in Fig. 11.1. The applications regarding the RFID implementation are designed with a separate GUI and appropriate local databases.

The application uses a centralized server that is a web, data, and processing server at the same time. It is equipped with Microsoft IIS Web server hosting the web application, Java execution environment, and Sun Spot SDK holding the main process for collecting the data and Microsoft SQLExpress as a data warehouse. The client accesses the service by using a web browser. The remote Sun SPOTs communicate with the Base Station connected to the server, which saves records in different databases according to the SPOT's ID and its relevant application. Those records are afterward used from the web application to visualize the data. Fig. 11.2 presents the start page of the web application's GUI (http://prosense.feit.ukim.edu.mk).

The complete solution comprises:

- Web-based ASP.NET web application consisting of flash animations, Ajax scripts, Cascading Style Sheets (CSS), and other tools for visualization.
- Database reader component part of the application used to read database records and display them depending on the scenario's requirements.
- Java daemon process running on J2SE and using Sun SPOT SDK that connects with the Base Station(s) and acquires the data, parses, and manipulates it and finally stores it in a database.
- Microsoft SQL Database, used to host records from the sensed environment.

The web application's GUI visualizes the five usage scenarios. Each scenario has a separate database and a different GUI (accessed through the start page) and can be

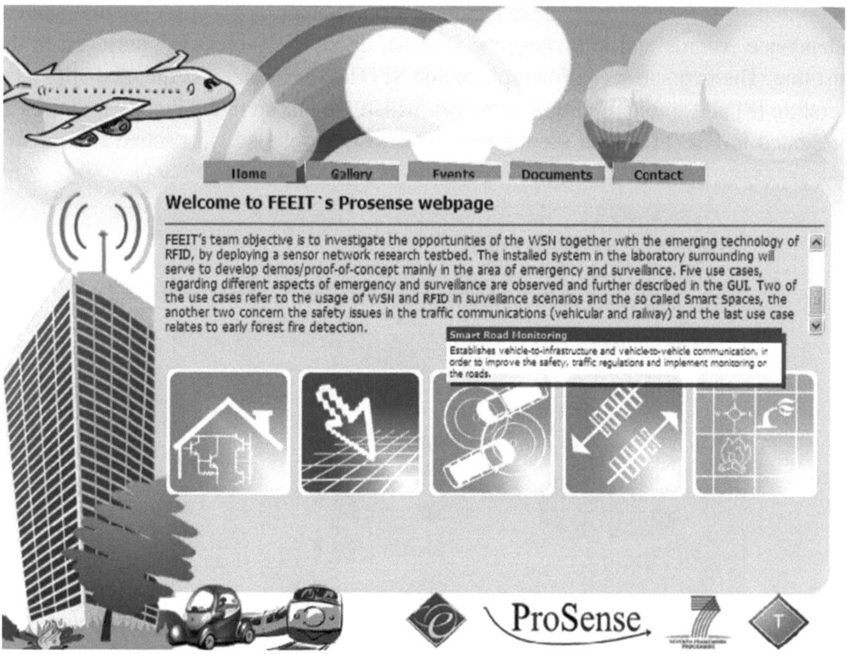

Fig. 11.2 FEEIT's GUI Start Page

used independently from the other scenarios. Each GUI briefly elaborates the usage scenario it refers to and presents the system architecture with the relevant real-time sensors' results. Additionally, some scenarios have the possibility to acquire statistics of the sensors' values variations over time. The RFID parts of the usage scenarios are developed as stand-alone applications and are included in the GUI with a short introduction. The separate usage scenarios are shortly elaborated in Section 11.3.

11.3 Testbed Realization

Each usage scenario is developed using specific parts of the testbed infrastructure. All scenarios can function as stand-alone applications, while at the same time they are integrated in the joint platform through the common GUI and the database. The features of the developed usage scenarios are shortly elaborated in this section.

11.3.1 Smart Institute

The *Smart Institute* scenario presents how the WSN, integrated with the emerging technology of RFID, is used to create a smart environment at the FEEIT's faculty premises. It comprises two separate demo applications, i.e., *Protected Institute*, which serves as an effective system for emergency situations' prevention (mainly protection of indoor fires) and *Presence Identification*, which is a stand-alone application for maintaining attendance records and informing the students about the professors' presence in the institute. The first sub scenario employs Sun SPOT WSN modules equipped with temperature [8] and smoke analog sensors [9], which are placed in the faculty premises as presented in Fig. 11.3. The second sub scenario is developed with active and passive

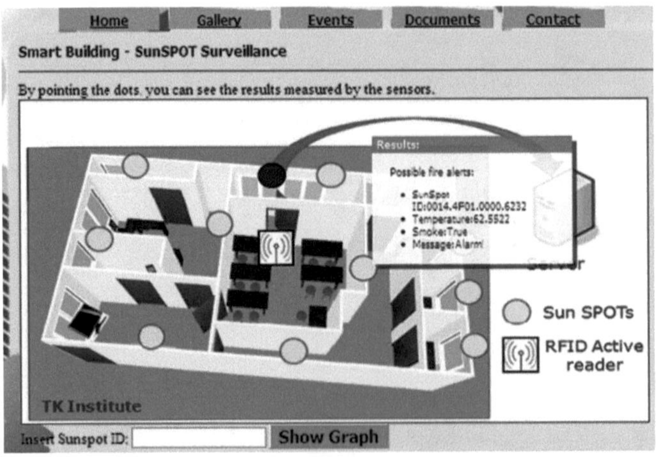

Fig. 11.3 FEEIT's premises with WSN and RFID reader location

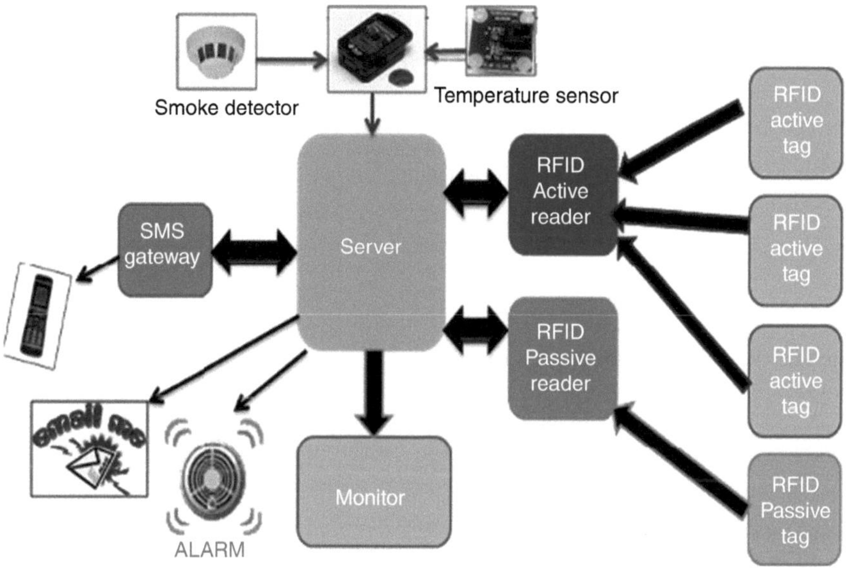

Fig. 11.4 Logical network architecture

RFID readers and tags [11, 13] and comprises a monitor placed at the faculty entrance where the students can check the presence of the professors they look for. The overall architecture is depicted in Fig. 11.4. The scenario is completed with a GSM/GPRS gateway [14], an alarm, and an e-mail sending capability of the servers' software for on-time notification in case of potential dangerous situation (e.g., fire).

The GUI (also presented in Fig. 11.3) displays each SPOT's ID with temperature and smoke values in real time and provides graphs for the values variations over time.

11.3.2 RFID Item Reminder

RFID Item Reminder is a usage scenario based on the Radio Frequency Identifier technology, which is commonly used for identification and tracking of objects and people. The scenario allows users to detect missing items and to ensure they have all their belongings when leaving homes or offices.

The *RFID Item Reminder* usage scenario is realized using passive RFID tags [12]. The system architecture is shown in Fig. 11.5. The application distinguishes two different types of demo applications, i.e., *permanent* and *non-permanent* demo application, enabling a 24/7 item observation and a one-time emergency use, respectively. The whole function of the system is enabled by a windows application on the host computer written in C-Sharp, which uses System.IO.Ports namespace from Visual Studio for making a connection to the serial RS232 port at which the reader is connected. The data with the tags' ID and the receiving time is stored in an ADO. NET database. In order to notify the persons about their missing items, the demo

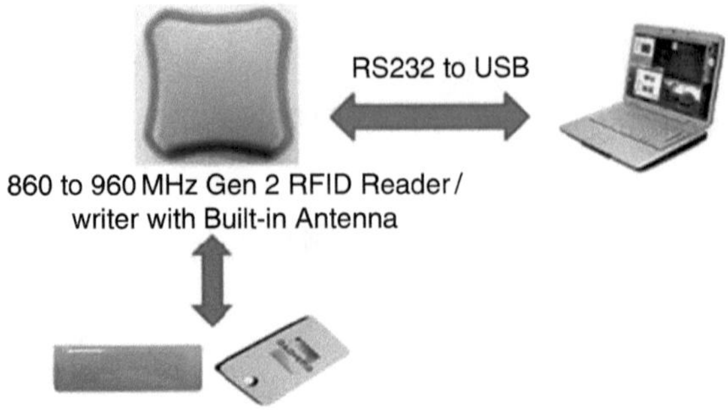

Fig. 11.5 RFID Item Reminder system architecture

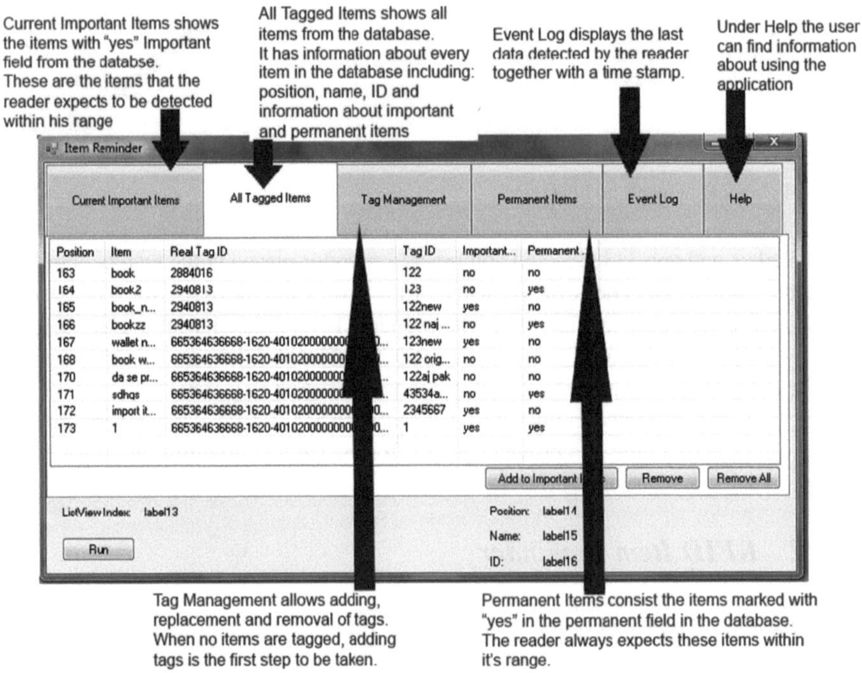

Fig. 11.6 RFID Item Reminder GUI

applications use a sound and an e-mail signalization. Figure 11.6 depicts the detailed description of the usage scenario through the GUI developed for this purpose.

11.3.3 *Smart Road Monitoring*

Smart Road Monitoring is a VSN application whose main idea is to aggregate safety, traffic regulation, and monitoring aspects, in order to improve traffic

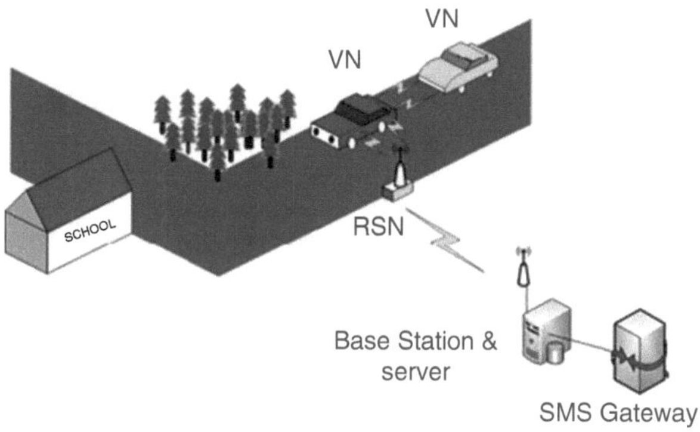

Fig. 11.7 Smart Road Monitoring network architecture

Fig. 11.8 VN and RSN components

regulation signalization and real-time reporting to the authorities. The small-scale setup is developed with toy vehicles and comprises Sun SPOTs with sensors placed inside the vehicle, representing the so called vehicle node (VN), a Sun SPOT roadside node (RSN) receiving the information from the vehicle sensors, a central server (data center) for gathering information from the RSN, and GSM/GPRS gateway for sending SMS messages (Fig. 11.7). Measurable features of importance for the scenario are the speed of the vehicle (measured with a tachometer placed on a toy wheel [2]), distance from other vehicles or objects (distance sensor placed on a toy front [3]), direction of movements (measured with Sun SPOT's accelerometer), etc.

The initial scenario is applicable in restricted areas such as schools, crossroads, and one-way roads, where the application is used to notify the drivers whether they are breaking the defined area rules. The GUI allows monitoring of the vehicles passing through the area used in the scenario (Fig. 11.8) and provides a graphical representation of the vehicle's speed over time.

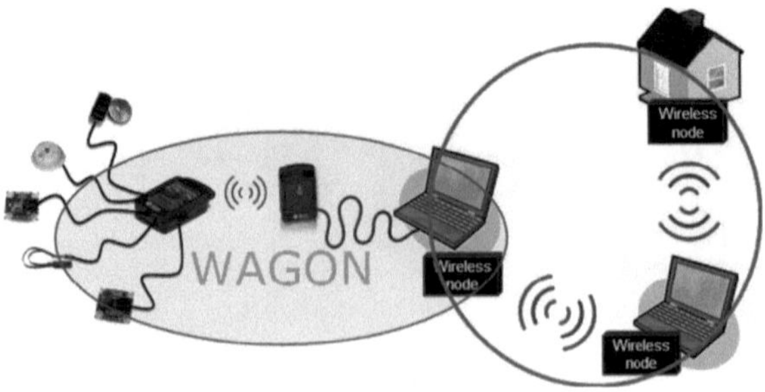

Fig. 11.9 Network architecture

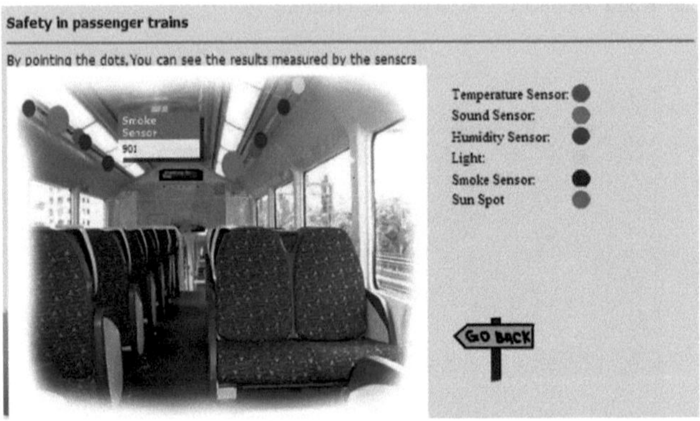

Fig. 11.10 GUI display of the sensors' values in passenger train

11.3.4 Railway Sensor Network: Secure Transport

The scenario for the *Railway Sensor Network: Secure Transport* is related to a monitoring system for railway vehicles. The system monitors the conditions of the critical components for the safety and comfort of the passengers as well as for the safety of the freight trains. The relevant parameters of interest are then wirelessly transmitted to the data acquisition center that can be placed either in the locomotive cab or in some ground station. The wagons' monitoring system comprises a wireless sensor unit (Sun SPOT) connecting various analog sensors (light [10], temperature and humidity [5], motion [4], sound [7], smoke [9]) that sense different parameters and transmit gained values to the Base Station. The BS is connected to a local host machine (PC) that retransmits the data through a particular wireless communication device to the distant database. The employed wireless devices are Xbee PRO rf modems (ZigBee [15]), HPS120 (Bluetooth [16]), and Avisaro (WLAN [17])). The overall network architecture is depicted in Fig. 11.9.

Figure 11.10 shows the scenario's GUI display with the real-time values of the sensors parameters deployed in the passenger train.

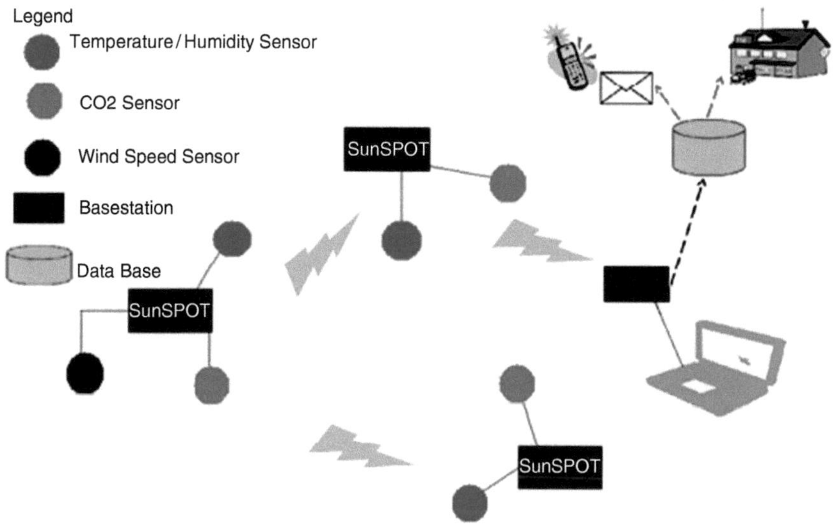

Fig. 11.11 Early Fire Detection network architecture

11.3.5 Early Fire Detection

Prevention and efficient monitoring of possible fires leading to disastrous consequences has become a global concern. The scenario called *Early Fire Detection* presents the idea and the design of a small-scale wireless sensor network testbed for early detection of fires mainly in outdoor settings (e.g., forest). The demo application is developed with WSN nodes, i.e., Sun SPOTs, different sensors (temperature and humidity [5], wind speed [6] and CO2 sensors [9]), a server and a GSM/GPRS gateway [14].

In order to achieve a small setup and to demonstrate the possibilities of the usage scenario, basic setups and algorithms are designed for three Sun SPOTs. Different algorithms in the SPOTs and on the server side serve to detect possible fire and its possibility of spreading and, also, to send notification messages. The network architecture of the basic demo platform is depicted in Fig. 11.11. Figure 11.12 shows the GUI display of the sensors' parameters.

11.4 Lessons Learned

Deployment of a WSN testbed and development of different demos are demanding tasks that require solving many issues. During the testbed setup, different problems were encountered mainly in the hardware implementation. The software implementation was also a challenging issue, but, as expected for Java programming, relatively easily solvable.

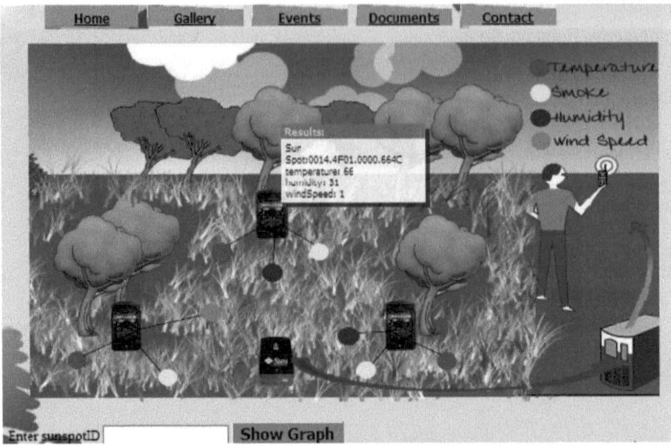

Fig. 11.12 GUI display of SPOT's parameters in Early Fire Detection

The main hardware difficulties are listed in Table 11.1. For example, the Sun SPOTs are suitable for software development and learning purposes, but inappropriate for development of real applications. The SPOTs have problems with frequent restarting and are interference sensitive, which makes them unsuitable for deployment in facilities with many devices operating on 2.4 GHz frequency band. Also, the output voltage they provide on their input/output pins is very unstable if the motes operate on battery. Therefore, the extra sensors that need to be attached on the Sun SPOTs should use a separate power supply (appropriate battery, for example) or the Sun SPOTs must be always connected to a power supply instead of operating on a battery. The rest of the equipment requires different and specific hardware connections, filters, wirings, or connectors. The comments for the different sensors and the equipment used in the mentioned scenarios are given in Table 11.1.

11.5 Conclusion

The presented testbed with the developed usage scenarios and demo applications comprises several aspects regarding the usage of WSN and RFID in emergency scenarios, surveillance systems, vehicular and smart home environments. The selected scenarios are implemented in a testbed environment with the purpose to show the possibilities of the technologies and to establish a basis for further research.

Even though the basic idea of the testbed is to demonstrate proof-of-concept of selected usage scenarios, it will serve as a laboratory where future research on various sensor networking aspects can be tested in a practical setup. They range from energy-efficient layer and protocol design, energy scavenging, energy harvesting to aspects of cooperative communications. Further on, novel and productive results can be also addressed in the fields of routing, attribute addressing, security, time synchronization, localization as well as efficient data gathering.

Table 11.1 Testbed hardware components

Equipment	Scenario	Comments about testbed hardware components
Sun SPOT [1]	1, 3, 4, and 5	– Problems with frequent restarting – Provides unstable output voltage – Problems for deploying an application 'over-the-air' (without connected Sun SPOTs on USB cable) – Interference sensitive
Temperature sensor – Phidgets SKU#33440 [8]	1	– Needs proper calibration – The temperature sensor component is rated at −50°C to +150°C, but the other components on the board, the connector and the cable, are rated at −30°C to +80°C
Smoke sensor MHz-11 [9]	1 and 5	– Highly sensitive – Needs proper calibration – Provides stabile digital and analog output
Speed sensor – tachometer SERVO TEK S-60 -P1104 [2]	3	– Specific wiring and connectors needed – Provides accurate results
Distance sensor – ultrasound distance sensor EZ0 [3]	3	– Problems when using the Sun SPOT battery – Requires separate batteries and specific connectors and wiring
Phidgets Precision Light Sensor [10]	4	– Needs proper calibration
Temperature and humidity sensor – Precon HS -2000V [5]	4 and 5	– Needs proper calibration (different sensors give diverse results)
Motion sensor – Phidgets SKU#33080 [4]	4	– Needs proper calibration – Sensitive to around 8m
Hitech Sound Sensor [7]	4	– Measures only above 80 dB
Wind speed and wind direction sensor A100L2 [6]	5	– Need to use universal board for connecting two wires from the anemometer to GND on the Sun SPOT (analog output is used) – Need to use condenser, like a filter, for more stable output voltage
Active Wave RFID kit [11]	1	– Incomplete version of the provided software – Asset tags encounter problems when entering into a local database
Anviz, OC200 RFID [13]	1	– Could not be connected by a network cable to PC, only by serial RS232 port – The reader data cannot be extracted into a separate database
GAO RFID reader and antenna [12]	2	– There is a minimum size limit for the tags – The reader needs an upgrade to be installed
GSM Module Cel-00757 and GSM Evaluation Board CEL-00477 [14]	1, 3, 4, and 5	– Working with AT commands via SerialPort – Using SMSLib library
XBee-Pro [15]	4	– Problem with power supply (work on adapter/battery)
HPS-120 [16]	4	– Works properly
WLAN-Box (W21133) [17]	4	– Requires establishing virtual WLAN connection

To sum up, the testbed is used for the development of the specified usage scenarios and will also serve for validation of research results and development of novel concepts and projects.

Acknowledgement The authors would like to thank Bisera Jankuloska, Milan Zahariev, Mihajlo Pavloski, Sanja Bonevska, Simona Temelkova, Natasa Paunkoska, Andrijana Jankova, Kiril Popadinski, Vanco Popov, Marijana Neskovska and Makedonka Gudzirova, undergraduate students in telecommunications (2008/2008) at the Faculty of Electrical Engineering and Information Technologies (FEEIT) – Skopje, for their valuable contributions and dedicated work on the usage scenarios within the ProSense project [18].

References

1. SUN Spot World, official Website for Sun SPOTs. Available at http://www.sunspotworld.com
2. Speed sensor, Tahometer, SERVO TEK. Available at http://www.servotek.com
3. Ultrasound distance sensor. Available at http://www.maxbotix.com/
4. PIR motion sensor, Sparkfun. Electronics website available at http://www.sparkfun.com
5. Precon relative humidity sensor. Available at http://www.preconusa.com/humidity_moisture_dew_sensors.htm
6. Wind Speed and wind direction sensor. http://www.windspeed.co.uk/ws/index.php?option=displaypage&op=page&Itemid=49
7. Hitech Sound Sensor. Available at http://www.robotshop.ca/hitec-sound-sensor.html
8. Phidgets Temperature Sensor SKU#33440. Available at http://www.hvwtech.com/products_view.asp?ProductID=260
9. MHZ-11 CO2 sensor. Available at http://www.hwsensor.com
10. Phidgets Precision Light Sensor. Available at http://www.phidgets.com/products.php?product_id=1127
11. Active Wave, RFID solutions. http://www.activewaveinc.com
12. GAO RFID. Available at http://www.gaorfid.com
13. Anviz Biometric Tech. Co., Ltd. Available at http://anviz.en.alibaba.com/search/product?IndexArea=product_en&SearchText=OC200
14. GSM/GPRS gateway – GSM Module Cel-00757 and GSM Evaluation Board CEL-00477, Sparkfun Electronics website. Available at http://www.sparkfun.com
15. XBee-pro rf modems. Available at http://www.adhocelectronics.com/XBP24-PKI-001-RA
16. HPS wireless modems. Available at 120 http://www.wcscnet.com/HdwHPS120.htm
17. Avisaro WLAN Box 2.0. Available at http://www.avisaro.com/tl/avisaro-wlan-box-20.html
18. ProSense, EU FP7 CSA project. Available at http://www.prosense-project.eu

Chapter 12
ETF WSN Testbed: Practical Health Enhancing Application of Wireless Sensor Networks

Zoran Babović, Aleksandar Crnjin, Goran Rakočević, Stanislava Stanković, and Veljko Milutinović

This chapter outlines the efforts of ETF Belgrade's team of students during the course of FP7 project PROSENSE. The project called upon ETF's young researchers and students to look into possible ways to use emerging sensor network technology for enhancement of personal and public health. ETF's researchers responded with a variety of proposed applications, ranging from a fitness monitoring application (Smart Running Track), to environmental hazard monitoring/avoidance (Interactive Street Sensing) and a middleware component for sensor network integration (Prosense Common Gateway).

12.1 The Smart Running Track

The Smart Running Track is a sensor network application targeted at personal health enhancement through observation of runners' body parameters (such as ECG and oxygen blood concentration) and fitness parameters (average and current speed, amount of calories burned) in order to provide feedback to runners and, optionally, trained advice from a physician.

Originally, the Smart Running Track was conceived as a three-part spatially distributed application, running on sensor nodes, mobile phones, and a desktop computer. Sensor nodes were supposed to communicate with mobile phones through Bluetooth radio and with a laptop computer through a gateway sensor node, via ZigBee radio. Unfortunately, the only sensor node technology to provide sensor nodes equipped with both Bluetooth and ZigBee radios – Shimmer Research's SHIMMERs – is designed with both radios sharing the same data bus, the consequence of which being the fact that the two radios cannot be used at the same time. Because of this limitation, and observed poor performance of Bluetooth connection in the presence of Wireless LAN and noticeably different performance when running as opposed to standing or walking, the SRT team has chosen to focus

Z. Babović (✉), A. Crnjin, G. Rakočević, S. Stanković, and V. Milutinović
University of Belgrade
e-mail: zbabovic@gmail.com

L. Gavrilovska et al. (eds.), *Application and Multidisciplinary Aspects of Wireless Sensor Networks*, Computer Communications and Networks,
DOI 10.1007/978-1-84996-510-1_12,

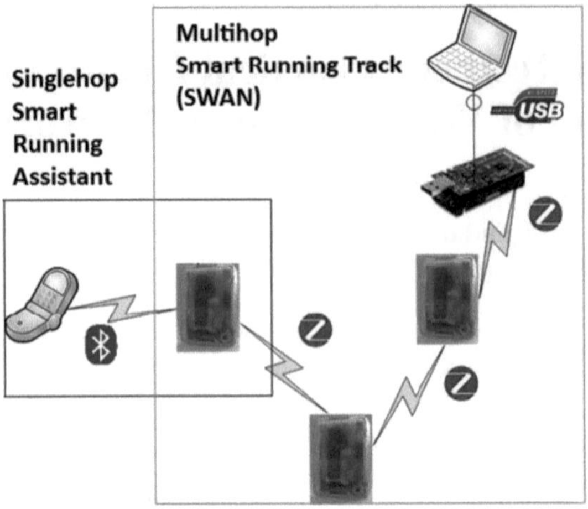

Fig. 12.1 Smart Running Track and Smart Running Assistant

on the second part of the scenario (involving sensor nodes and a laptop computer). The first part (interconnection between a mobile phone and a sensor node) has been since treated as a spin-off project – the Smart Running Assistant (Fig. 12.1).

12.1.1 SRT Hardware

The hardware for Smart Running Track includes the following devices:

- Shimmer Research SHIMMERs, some of them additionally equipped with Shimmer Research's ECG and pulse oximetry sensing equipment (Fig. 12.2). Typically, a runner wears up to three SHIMMER nodes – one of which records hip movements, one records ECG signals.
- TelosB nodes, one per Smart Running Track deployment. A TelosB is plugged into a laptop computer through its USB port, serving as a gateway between a sensor network and the desktop Smart Running Track application.
- A laptop computer running the SWAN (Sensor Wireless AcquisitioN) software, which collects and processes the sensor data.

12.1.2 SRT Software

The software for the Smart Running Track is centered around a Java application running on a desktop computer called the SWAN (Sensor Wireless Acquisition) GUI. The application was developed using the NetBeans integrated environment and the Swing API (for user interface; Fig. 12.3).

When plugged into the USB port, a TelosB node registers itself as a serial port (usually COM15 or COM16) through a USB serial driver. The SWAN can thus use

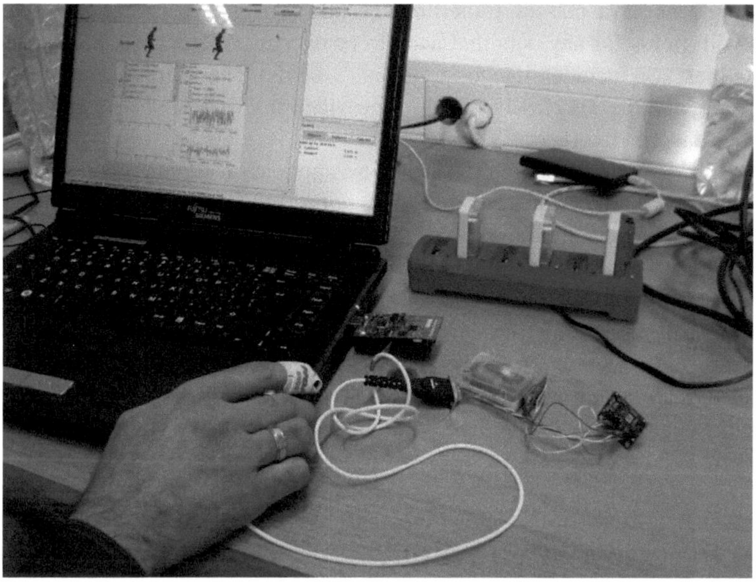

Fig. 12.2 Pulse oximetry on SHIMMERs in a lab setting

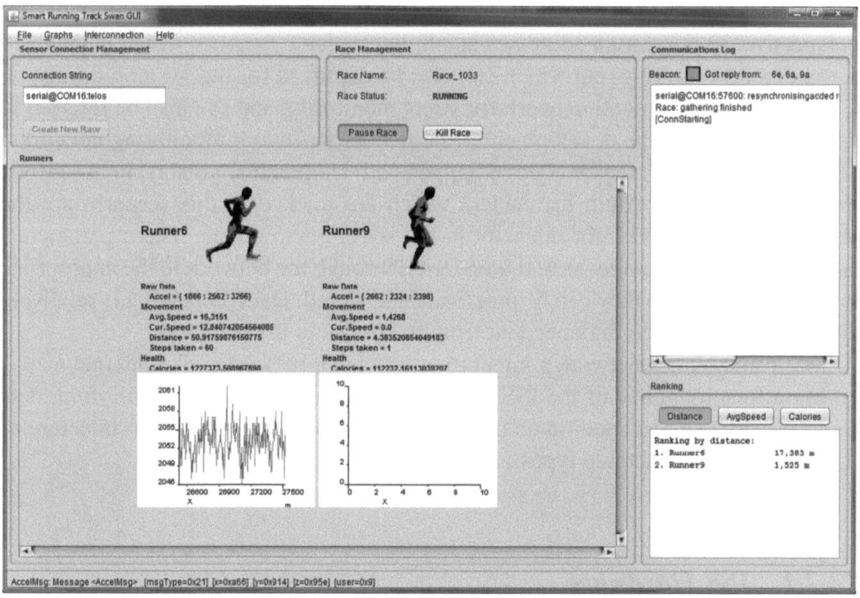

Fig. 12.3 Screenshot of the SRT Sensor Wireless Acquisition GUI

the Java Communication API (JavaComm) to send and receive raw bits to and from the TelosB node. However, instead of sending/receiving raw bits, TinyOS provides Java classes which streamline the process and enable the persistence of the connection as well as the exchange of structured Active Messages. Active Messages are represented as Java classes (inheritors of the basic net.tinyos.message.Message class) with fields for any data that need to be exchanged (for example, an accelerometer message includes integer x, y, and z fields) Therefore, a combination of JavaComm and the TinyOS Java classes provides an efficient way to communicate with the sensor network.

The TelosB node is running the TOSBase application supplied with the TinyOS distribution. The sole purpose of this application is to relay messages coming from the sensor network and addressed to the sink, to the desktop application that has registered itself as a listener; and to relay messages coming from the desktop application to specific nodes in the sensor network to which they are addressed.

SHIMMER nodes are running either of the three applications: AccelSense, ECGSense, or OxySense. These are TinyOS applications with the purpose of utilizing the onboard A/D converter and DMA controller to collect measurements from respective analog sensors and then send these measurements via radio.

12.2 Interactive Street Sensing

Interactive Street Sensing (ISS) network is a wireless sensor network which offers new approach to walking activity from medical as well as from human daily-life perspective. This solution addresses small, but scalable network which is intended for outdoor monitoring, but which can also be deployed for indoor monitoring.

ISS network is a small network for street monitoring and can also be referred to as a framework for the development of large outdoor/indoor monitoring networks.

This network uses sensor nodes equipped with temperature sensors, light sensors, and accelerometers. Main parameters which are tracked in this scenario are the temperature level and light intensity in the street environment. These parameters are sampled from the sensors and transferred through the network to the appropriate host machine, where they can be processed and stored. Data processing is relatively simple, but completely independent from the nodes in the network.

ISS also assumes network's integration into large networks, like Internet.

Main functionalities of the ISS network are street parameters monitoring, data transfer, and data presentation on the Internet (Fig. 12.4). Complete architecture of the network includes three types of devices (Fig. 12.3).

12.2.1 ISS Hardware

ISS network architecture (Fig. 12.3) contains three conceptual parts:

- Sensor nodes
- Base station
- Server

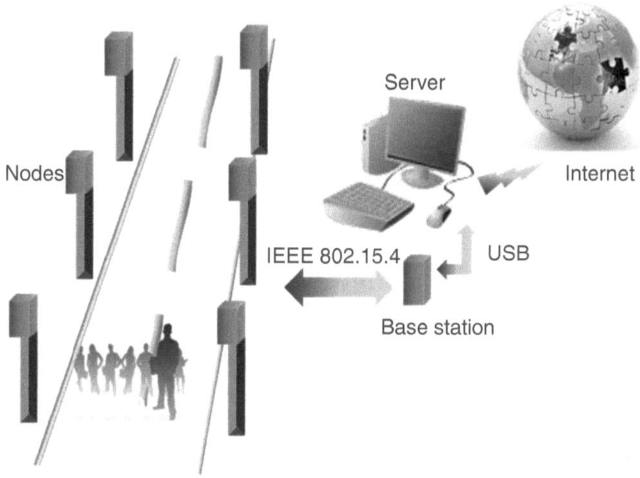

Fig. 12.4 Interactive Street Sensing (ISS) architecture

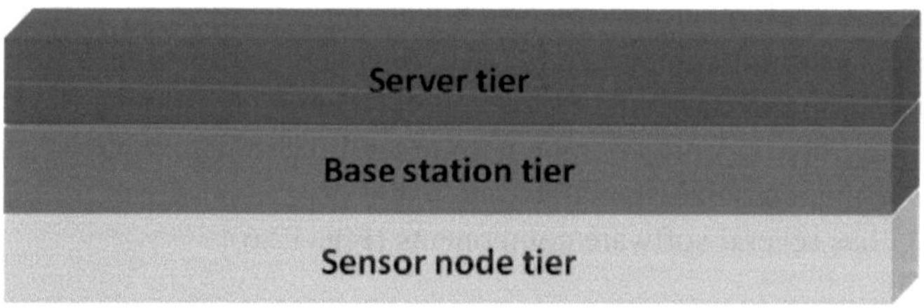

Fig. 12.5 ISS software tiers

Sensor nodes are Sun SPOT devices equipped with a number of sensors. These nodes are capable of gathering relevant sensor data from the environment and transferring the information to the base station (via 802.15.4 link).

The base station, a Sun SPOT without the sensor board, is attached to server (host machine) through USB connection. This device is used for data transfer from nodes to the server. The server component maintains a database in which collected information is stored, offers an appropriate graphical interface, and integrates the whole ISS sensor infrastructure with Internet.

12.2.2 ISS Software

ISS software is divided into three different tiers (Fig. 12.5):

- Sensor node tier
- Base station tier
- Server tier

The first tier is responsible for sensor data sampling and sensor data transfer to the base station. The second tier includes software for communication to sensor nodes and data transfer to server. The third one is deployed on the server and has two important services, the database management and the graphical data representation.

12.3 ProSense Common Gateway

Starting as an effort to integrate all Personal Body Area Networks with purpose of personal health care, the Prosense Common Gateway (PCG) has become a middleware software component with aim to offer an infrastructure for integrating various wireless sensor networks to one global sensor network. The main idea is to create a layer between user applications and various wireless sensor networks, and to enable to those applications access to sensor data deployed everywhere.

It is planned that PCG should sit at desktop computer which is connected with various WSN or with other devices which are connected directly with WSN. PCG provides users a common interface for searching and gathering data from that unified sensor data space. Basically, PCG supports recommendations from OCG SWE working group, but it is not compliant completely. Users are able to search for a sensor data using temporal, spatial, or any value-dependant filters.

The PCG has several software components (Fig. 12.6):

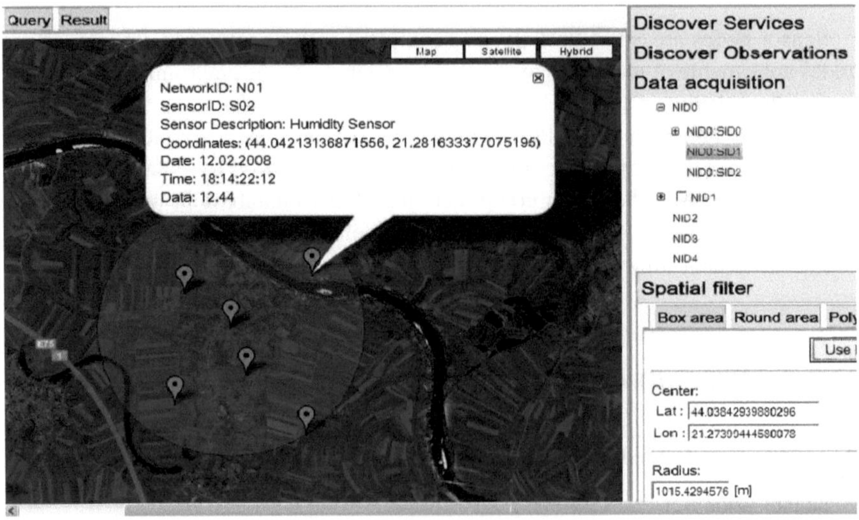

Fig. 12.6 Graphical Interface for using Prosense Common Gateway

- The Core PCG is a middleware server component that publishes interface(web interface) to the outside world, and contains all logic for managing and searching gathered sensor data. It is connected with appropriate database in the background and usually runs on application server.
- The PCG demonstration application is a user-located web application, which is used as an example of how to use the PCG and to utilize potentials of the PCG. It is developed as a web application, with integration of popular geo-maps (GoogleMaps).
- The Client PCG component, which is a part of client WSN desktop component, and is responsible for communication with the Core PCG. There is a special interface for interaction between WSN applications and Core PCG.

In order to store and manage data from various sensors, the PCG uses a meta-model suitable for this purpose. Meta-sensor defines all possible data that could be contained in related sensor type, and every value is defined by appropriate meta-property. As a consequence, each sensor in the system is defined by appropriate meta-sensor, which defines what properties (value types) could have that sensor, or what properties (defined by meta-property) could be changed during time. For example, we can define meta-sensor of type BasicTemperatureSensor, which can have the following properties (value types): temperature value, unit of that temperature value (°C, °F, K, ...), precision (which could be defined as constant during all measurements, or could vary from one measurement to another), location (geo-location of the sensor, which could be constant – sensor is static, or could change during time – sensor is mobile). The obligatory values for all sensor types are flags if sensor is active or not (does it work or not), and if sensor is online or not (could be message routed to it). The list of all meta-sensors and meta-properties (meta-definitions in PCG terms) could be extended during time.

At the beginning of integration with the PCG, the sensor network should be registered to the PCG. A sensor network is defined by its URL, name (internal PCG name), and by its owner. After registration of the sensor network, it could be registered sensors deployed in that network. For every new sensor its sensor type must be defined (by meta-sensor described above), as also its unique internal ID inside the sensor network. Depending on the selected sensor type, various properties will be added to that sensor, and could be potentially populated at the registration time. Since sensors are registered, their observations are allowed to be registered (measurements) in time. Registration of the sensor network must be done through the PCG management application, but registration of sensors could be done either through management application, or directly through client PCG component (which is part of some other application dealing with appropriate sensor network). Registration of sensor observations is possible only through client PCG communication interface.

For enabling an easy integration with the PCG Core, the client package is provided with appropriate classes developed in Java. This package contains classes for connecting to the PCG Core server through socket communication, which is used for bidirectional communication between the PCG core and client application.

Usually, at the beginning of communication the client sends to the PCG server a login request, and after receiving login confirmation requests for getting meta-definitions, and sensor network object where this client belongs. After this initial message exchange, the client is ready to send either requests for registration of new sensors, or to notify the PCG server with the latest sensor observations. The client PCG package contains some utilities classes that facilitate creating of appropriate sensor observation transfer object for a certain sensor. This sensor observation transfer object is needed for notifying the PCG server with latest observations from sensors.

All Internet users are potential users of PCG service. A demonstration application is developed (see Fig. 12.1 above) for showing possible modes of using PCG capabilities. There are several use cases. User can manually discover which sensor networks are connected to the PCG. For each sensor network, there is a list of sensors deployed in that sensor network. If location property for those sensors is provided, users can see on the map where that sensor is deployed by clicking on the sensor in the list. Another possibility for users is to search for some sensor data using temporal, spatial, and values filters. Temporal filters apply on sensor observation, with the possibility of selecting observation in some time interval, after some time, before some time, or just latest gathered observations. Spatial filters are used for selecting sensors from some area. This area could be determined using the map, and there is an option of box area, or polygonomial area. Values filters are offered if user selects searching by some sensor type. Based on appropriate meta-sensor, several properties could be included in search criteria, as a condition with desired values that observations from sensors must satisfy. After submitting queries to the PCG, user receives response as a list of sensor observations or sensors that satisfy the query. Gathered sensor observations could be browsed using the map or using the list. User can also get history of sensor observations for selected sensor. Additionally, user can subscribe for receiving the latest observation for selected sensor and automatically display it using the table or graph (Fig. 12.7).

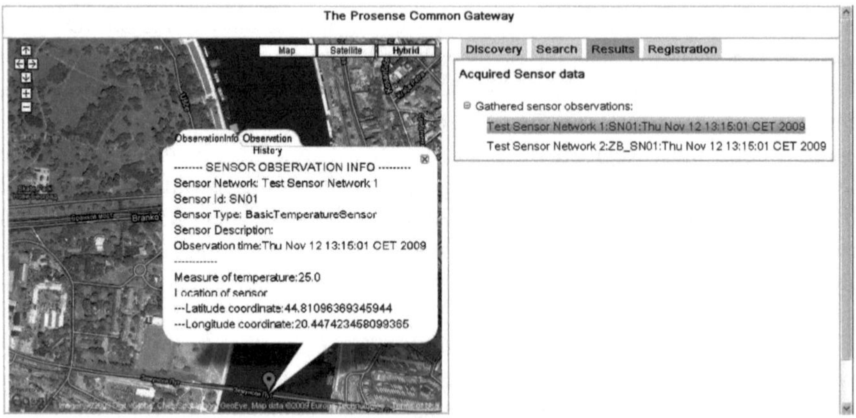

Fig. 12.7 Sensor data acquistion through the PCG

Part IV
Birds of a Feather

Ubiquitous Sensor Networks
eMuseum
The PAR Logistics Solver Problem
Wireless Sensor-Based Robot Control

Chapter 13
Ubiquitous Sensor Networks

Zhilbert Tafa

Ubiquitous computing is a new concept that could be defined as a technology which can be available anywhere that it is useful and economically viable to expect to find a sensor. Sensor network is recognized to be a key technology for building a ubiquitous system. Small physical dimensions of sensor nodes are often one of the musts in a typical ubiquitous network, since it is dedicated to unobtrusive integration into living, working, scientific, industrial, and other environments. This implies high integration of sensing, computing, and communication capabilities of the devices meeting at the same time application-specific demands. Using mostly wireless infrastructure, ubiquitous technologies are supposed to interface the physical environment in various circumstances such as military, home, working, health, industry, etc. This makes implementation of the ubiquitous sensing highly application-specific. Interfacing issues along with the interoperability between different technologies are the most demanding parts of the system design. Many constraints here bring up challenges in various engineering fields such as sensing, energy conservation, communication, security issues, etc. These issues are often complementary coupled, which makes the process of finding the optimal solution more complicated.

The main components of a ubiquitous sensor network (USN) can be grouped in [1]:

- **Sensor Network**: Comprising sensors, independently energy supplied, used for collecting and transmitting information about their surrounding environment
- **Access Network**: Intermediary or sink nodes collecting information from group of sensors and facilitating communication with control center or with external entities
- **Network infrastructure**: Based on future NGN (next-generation network)
- **USN Middleware**: Software for the collection and processing of large volumes of data
- **USN Applications Platform**: A technology platform to enable the effective use of a USN in a particular application

Z. Tafa (✉)
Montenegro Telekom
e-mail: Zilbert.Tafa@telekom.me

L. Gavrilovska et al. (eds.), *Application and Multidisciplinary Aspects of Wireless Sensor Networks*, Computer Communications and Networks,
DOI 10.1007/978-1-84996-510-1_13,

Because of the great number of constraints and the application field specifics, the main USN challenges today are mostly concentrated on sensor network itself. These issues and challenges can be grouped in [2]:

- **Autonomous setup and maintenance**: Sensor nodes are randomly deployed and need to be maintained without any human intervention
- **Infrastructure-less**: All routing and maintenance algorithms need to be distributed
- **Energy-conscious design**: The microcontroller, Operating System, communication protocols, and application software should be designed to conserve power
- **Global time synchronization**: Sensor nodes need to synchronize with each other in a completely distributed manner for communication synchronization, temporal ordering of detected events, and elimination of redundant events/messages
- **Dynamic topology due to failures or power-down/up**
- **Real-time and secure communication**

Along with the tendency of minimization in physical size of sensor nodes and other application-specific constraints, the given aims should more strictly be followed only in limited number of the very demanding applications (such as battlefield surveillance and monitoring, etc.). However, most of the typical implementations today are deployed using small-scale ad hoc networks based on tree or cluster tree topologies where some of the nodes do not need to be battery supplied. Within this context, many ubiquitous industrial, environmental, healthcare, and other systems are already widely deployed [3, 4].

References

1. ITU-T Technology Watch Breifing Report Series (2008) Ubiquitous sensor networks, No.4, Feb 2008
2. Choi L (2007) Ubiquitous networks: sensor networks, computer system lab. Fall
3. Katsumi T, Kenji K (2006) Ubiquitous sensor network system. NEC Tech J 1:78–82
4. Mukhopadhyay S, Gaddam A, Gupota G (2008) Wireless sensors for home care monitoring – a review. Bentham Science Publishers Ltd. pp 32–39

Chapter 14
eMuseum

Aleksandar Ćorović, Bojan Imamović, Edin Kadrić, Lejla Kadrić, Nermin Lipa, and Selma Opanović

14.1 Introduction

The main idea of this project was to implement WSNs [1] in bringing new content to museum visitors, making the visit more interesting and, thus, attracting more visitors. We created two applications: security application – which uses sensor abilities to monitor exhibit surroundings, and visitor's application – which communicates [2] with exhibit sensors and identifies them. Each exhibit is equipped with SunSPOT device, which continuously broadcasts its address for visitor's application to pick up and on regular time-cycle broadcast sensor readings for security application. Each visitor would receive PDA type of device, which would after identifying the exhibit, display content for it. The content could be text, audio, video, or images, depending on the exhibit. Visitor application also displays current position on the museum map. Map can be used to guide the visitor through the museum, either by standard guide tours or if visitor wishes to see certain exhibits. Exhibit sensors monitor light, temperature, and acceleration. This information is broadcasted after fixed period of time (if every reading is within given range) or immediately, if some of the readings go over the given threshold. Security officer is presented with the map of the museum, where every exhibit is displayed with light, temperature, and acceleration readings.

14.2 Current status

At the moment of writing (November 2009), we created both applications – visitor's and security. They are operational and core functionalities are present. We based our work on two types of code: Java and C#. Java code is deployed

A. Ćorović (✉), B. Imamović, E. Kadrić, L. Kadrić, N. Lipa, and S. Opanović
University of Sarajevo
e-mail: peglaboy@gmail.com

L. Gavrilovska et al. (eds.), *Application and Multidisciplinary Aspects of Wireless Sensor Networks*, Computer Communications and Networks,
DOI 10.1007/978-1-84996-510-1_14, © Springer-Verlag London Limited 2011

on SunSPOTs [3], [4] and that part of the project is finished. Visitor's and security applications are written in C#. Basic functionalities are present. In visitor's application, exhibits are recognized and content is displayed. Position is displayed on map. In security application, exhibit and all readings are displayed. If some of the readings go over the given threshold, alarm is displayed.

14.3 Further Development

Further development has to go in two directions – improvement of sensor's Java code and improvement of applications.

Sensor code has to be optimized to support many visitors. Broadcasts for security application have to be scheduled, so problems with many concurrent broadcasts could be avoided.

Further development for visitor's application include adding guided tours, recommendations based on already seen exhibits, statistics of visited exhibits so their position could be improved, adding support for avoiding bottlenecks.

Further developments for security application include museum automation. Based on sensor readings, application should turn lights on/off, turn AC on/off. In case of acceleration alarm it should close certain parts of museum.

14.4 Conclusion

There are many advantages of implementing this idea, which are stated in the introduction. Disadvantages include high initial cost of deployment, high possibility of malfunctions caused by improper handling, and time of implementation.

The great thing about this idea is that this concept is applicable to many areas: all kinds of exhibits, fairs, Zoos, and others which include bringing content to the visitor.

References

1. http://en.wikipedia.org/wiki/WSN
2. https://www.sunspotworld.com/forums/
3. SunSPOT Developer's Guide, SUN Microsystems (2009)
4. http://www.sun.com/

Chapter 15
The PAR Logistics Solver Problem

Gordana Nikolić, Dario Zorić, Alan Martinović, and David Dubrović

15.1 Introduction

Business academy Rijeka is a lifelong learning academy. Because of the constant, dynamic evaluations in economy, nowadays lifelong learning is a useful premise. Our partner in this project was 4LOOK. The company 4LOOK represents international companies, imports and sells professional equipment and tools of the highest quality for hair stylists and beauty salons, encompassing by its width all requirements of our buyers. 4LOOK products are widely recognized with superior quality, design, and application at the top of world technology. The main goal of the logistics and distribution is to provide a time and space optimal, and cost-effective flow of materials, goods, and information in the company.

We propose creating an infrastructure whose plug-ins could easily be implemented. Application would be designed to plan and measure the optimal route between locations.

15.2 Infrastructure and Application

The infrastructure includes SunSPOT devices placed on all major crossroads for purposes of collecting time and additional crossroad information. They should be powered either through the traffic light power supply or by an external source. A device should also be placed on each delivery vehicle and serve as a storage device and a communicator. A SunSPOT base-station would be placed in the company headquarters' parking place. The spot should be chosen carefully so that the transmitting area of the base-station covers all the parking places. The base-station would be connected to a desktop computer that contains a functional database.

G. Nikolić (✉), D. Zorić, A. Martinović, and D. Dubrović
Business Academy Rijeka
e-mail: gnikolic@par.hr

L. Gavrilovska et al. (eds.), *Application and Multidisciplinary Aspects of Wireless Sensor Networks*, Computer Communications and Networks,
DOI 10.1007/978-1-84996-510-1_15,

Wireless sensor networks (WSN) built from huge amounts of interacting nodes is the basis for this research [1]. Issues such as mobility, network size, deployment density, and energy are key factors for the development of new communication methodologies.

Daily route planning is made with an application for tactical planning of transport, with the help of processed data collected through previous deliveries. As time passes, the data collected throughout daily deliveries are statistically processed, and they provide a base for determining optimal routes in the future.

15.3 Realization

As the delivery vehicle starts its daily shift, it leaves the company parking. It also receives the data from the base-station about the time of departure. While it delivers the goods throughout the day, it passes through different crossroads. When the vehicle steps into the emitting area of the crossroad spot, it receives and stores the information about the crossroad it just entered and the current time. The time when the vehicle leaves the crossroad is also stored.

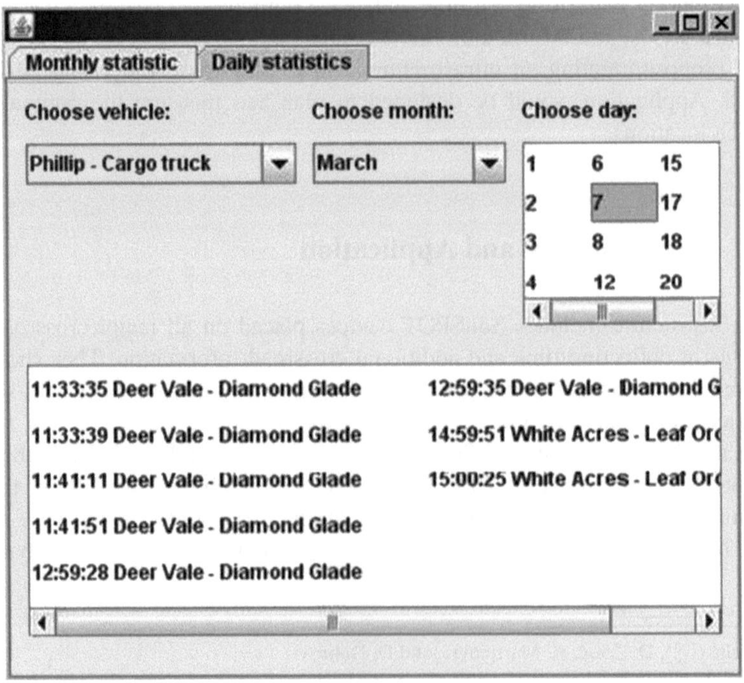

Daily Statistics

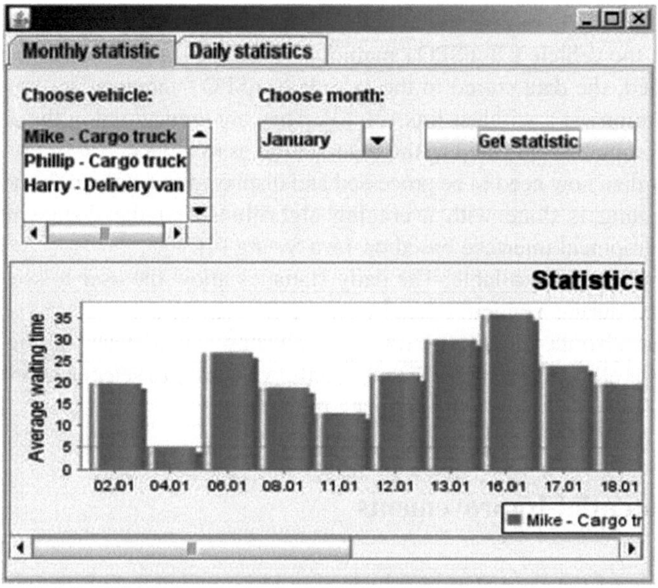

Monthly Statistics – Single Selection

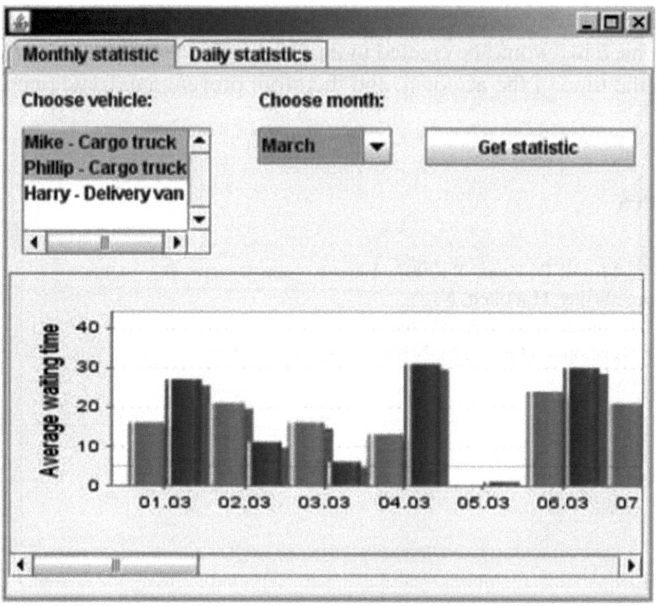

Monthly Statistics – Multiple Selection

When the vehicle returns to the base, all the data about the crossroads it visited is stored in the vehicle's SunSPOT memory. When the signal from the base-station is recognized, the data stored in the vehicle SunSPOT memory are transferred to the base-station via a wireless link [2], [3]. They are then stored in the database on the local computer connected to the base-station as raw data.

The raw data now need to be processed and displayed in a human-friendly format. The processing is done with averaging algorithms and the displaying is done through a graphical interface based on Java Swing [4].

Two displays are available. The daily statistics allow the user to see the crossroads visited during a specific date for a specific vehicle.

There are also the monthly statistics which show the average waiting time on a crossroad for all the days of a selected month. An option to select multiple vehicles is also available so the user can compare the times.

15.4 Potential Improvements

1. The logic of the base-station-vehicle should be updated with the client–server type of connection to prevent conflicts of datagrams.
2. The accelerometer could be used for additional route analysis. It would measure the number of starting and stopping of the vehicle. This information would be useful since it can be connected to the fuel consumption and therefore help in determining the optimal route.
3. Additional use of the accelerometer could be in accident diagnostics. When an extreme acceleration occurs, one that the vehicle could not achieve in regular conditions, a log would be created in the SunSPOT memory. The log would then contain the time of the accident, and therefore prevent any fraud opportunities.

References

1. Sohraby K, Minoli D, Znati T (2007) Wireless sensor networks: technology, protocols, and applications. Wiley, Hoboken, NJ
2. https://www.sunspotworld.com/forums/
3. SunSPOT Developer's Guide, SUN Microsystems (2009)
4. http://www.sun.com/

Chapter 16
Wireless Sensor-Based Robot Control

Gyula Mester, Istvan Matijevics, Tamas Szepe, and Janos Simon

16.1 Introduction

The chapter deals with the wireless sensor-based remote control of mobile robots motion in an unknown environment with obstacles, using the Sun SPOT technology. The proposed method has been implemented on the miniature mobile robot Khepera that is equipped with sensors and the free range Spot. The chapter includes the implementation of mobile measuring station in greenhouse environment.

16.2 Wireless Sensor-Based Remote Control of Mobile Robots

In this chapter we have used SunSPOT-s (Small Programmable Object Technology) to achieve remote control over a Khepera® mobile robot. When the vehicle is moving toward the target and the sensors detect an obstacle, an avoiding strategy is necessary [1]. The Sun SPOT connection strategy is presented in Fig. 16.1. The experimental results show the effectiveness and the validity of the obstacle avoidance behavior in an unknown environment of the proposed control strategy.

16.3 Mobile Robot Measuring Station in Greenhouse Environment

The applications for WSNs are many and varied. Using autonomous measuring systems (Fig. 16.2) helps to monitor all the necessary parameters for creating the optimal environment in the greenhouse [2]. The robot equipped with sensors is capable of driving to the end and back along crop rows inside the greenhouse.

G. Mester (✉), I. Matijevics, T. Szepe, and J. Simon
University of Szeged
e-mail: gmester@inf.u-szeged.hu

L. Gavrilovska et al. (eds.), *Application and Multidisciplinary Aspects of Wireless Sensor Networks*, Computer Communications and Networks,
DOI 10.1007/978-1-84996-510-1_16, © Springer-Verlag London Limited 2011

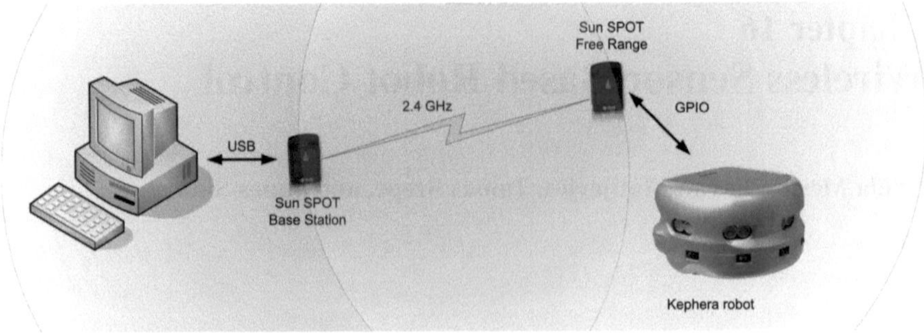

Fig. 16.1 Remote control system

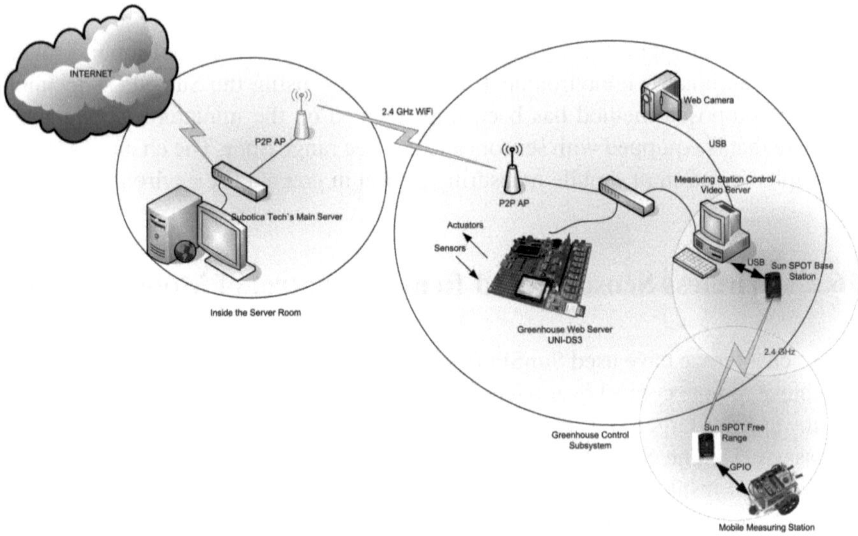

Fig. 16.2 Greenhouse control system

16.4 Conclusion

The chapter deals with the wireless sensor-based remote control of mobile robots in an unknown environment with obstacles, using the Sun SPOT-s. The chapter includes the implementation of mobile measuring station in greenhouse environment.

Acknowledgment This research was partially supported by the TÁMOP-4.2.2/08/1/2008-0008 program of the Hungarian National Development Agency.

References

1. Mester G (2009) Intelligent mobil robot control in unknown environments. In: Intelligent engineering systems and computational cybernetics, Springer, London, pp 15–26
2. Matijevics I (2009) Wireless sensors networks – theory and practice. In: Towards intelligent engineering and information technology, Springer, London, pp 405–417

Index

MIX
Papier aus verantwortungsvollen Quellen
Paper from responsible sources
FSC® C105338

If you have any concerns about our products,
you can contact us on
ProductSafety@springernature.com

In case Publisher is established outside the EU,
the EU authorized representative is:
Springer Nature Customer Service Center GmbH
Europaplatz 3, 69115 Heidelberg, Germany

Printed by Libri Plureos GmbH
in Hamburg, Germany